The Theory of Multidimensional Reality

By

Douglas B. Vogt

Member of the Geological Society of America

Research Funded by
The Diehold Foundation, Inc.

Published by
Vector Associates
JACKSONVILLE, FLORIDA
2015

VECTOR ASSOCIATES
PO Box 19792
Jacksonville, Florida 32245

First Printing 2015
10 9 8 7 6 5 4 3 2 1

Library of Congress Cataloging in Publication Data

Names: Vogt, Douglas B., 1947- author.
Title: The theory of multidimensional reality / by Douglas B. Vogt, Member of the Geological Society of America.
Description: Jacksonville, Florida : Vector Associates, 2015. | 2015 |
   Includes bibliographical references
Identifiers: LCCN 2015044520 (print) | LCCN 2015045690 (ebook) | ISBN 9780930808105 (alk. paper) | ISBN 093080810X (alk. paper) | ISBN 9780930808112 ()

Subjects: LCSH: Science--Philosophy. | Reality. | String models. | Quantum theory.
Classification: LCC Q175 .V6384 2015 | DDC 111--dc23 LC record available at http://lccn.loc.gov/2015044520

ISDN 978-0-930808-10-5

Table of Contents

Chapter 3

Chapter 4

Dedication

To

I had to think hard about who I would give the most credit to my ability to prove this information theory of existence. I came up with the following scientists: Dr. Thomas Gold, Dr. Bart J. Bok and the hundreds of other astronomers, astrophysicists, mathematicians who work tireless hours late at night trying to measure the distances, location and output of the stars in our Galaxy. For their great work in producing catalogs of the stars and trying to make sense of the great mystery, they see in the heavens. Without their work, I would have had a much harder time trying to prove the Theory of Multidimensional Reality.

Preface

This book is a pure philosophy of science book. The book presents an information theory of existence that is totally different from what you were taught in college or universities. Earlier versions of the theory were presented in three of my previous books, which covered many other subjects besides science philosophy. The first book *Reality Revealed, The Theory of Multidimensional Reality* dates back to 1977. That book covered the beginning of this theory/philosophy and applied it to some of the phonemes in science. Other authors have "borrowed" this theory of existence and applied it to various subjects. Some authors just changed the name but tried to describe the same thing, but it was obvious they did not understand it. Motion pictures also used parts of the Theory of Multidimensional Reality in various ways. It appears I am the first to describe and apply an information theory of existence to many of the toughest phenomena in the Universe. This current book represents about 45 years of research in most of the hard sciences including astronomy, geophysics, electronics, physics, etc. In my first book, I even had a chapter on psychic phenomena, by which I explained many of the paranormal phenomena using this information theory of existence. Even though there was only one chapter out of twelve that was dedicated to explaining parapsychology, the Library of Congress originally indexed the book Psychical research, which I could never understand. In mid April of 1994, I had the opportunity to visit the Library of Congress in Washington D.C. I asked them why they indexed it that way and requested the classification be changed. Much to my surprise they told me that one of the intelligence agencies in the U.S. Government reclassified my book on April 4th just two weeks before, to Philosophy of Science (Q175.V638 1978).

The Theory of Multidimensional Reality answers the three most difficult questions in the Universe: Why was the Universe created? If God created the Universe then who created God? And finally the big one: What is man's purpose in the Universe? This philosophy is the only one that can logically answer all three of these eternal questions.

Some of you may think that a science philosophy book is irrelevant to your everyday life. You cannot be more wrong. The philosophy presented is what I call a *Foundation Philosophy*. Such a philosophy is one that everything, including all science observations and experiments, are built upon and filtered through. All philosophies are derived from a foundation philosophy. Currently, society's foundation philosophy is a matter-oriented theory of the Universe thanks to Aristotle and the Church. You will learn where man made a terrible mistake. Such a philosophy prevents man from discovering how the Universe works and that there are cycles in time that almost completely destroys everything we have created including man.

"Let no one be slow to seek wisdom when he is young, nor weary in the search of it when he has grown old. For no age is too early or too late for the benefits of Philosophy." Epicurus, Letter to Menoeceus

Chapter 1
Defining the Problem

There are only two ways to explain how the Universe works. The currently accepted teaching is that matter is the dominant thing in the Universe, and everything could be explained by understanding the relationship between matter and energy. The only alternative explanation is that the Universe is the product of information. The information creates the matter world we live in, and it is transmitted from another time-space relationship, into our time and space to create our reality. Only one philosophy of these two philosophies can be correct.

Some thirty years after my first book *Reality Revealed, the Theory of Multidimensional Reality* (1977), I expanded and refined the theory in my third book *God's Day of Judgment, the Real Cause of Global Warming* (2007). The main purpose of both books was to explain what caused the geomagnetic reversals, the ice ages, mass extinctions and why they occur together. The Theory of Multidimensional Reality is used as the foundation philosophy that explains all of the phenomena associated with these events and why they occur cyclically through time. It has now been 39 years since I first presented this information theory of existence. I felt it was time to dedicate a book covering only the Theory of Multidimensional Reality without the distractions of other subjects. The book will cover how the theory explains all the major building blocks of our Universe such as magnetism, gravity, light, mass, and time.

First let us define what is philosophy. The dictionary definition is: The search for truth wisdom or knowledge through logical reasoning rather than factual observation, an analysis of the grounds of and concepts expressing fundamental beliefs. Philosophy is the theory or logical analysis of the principles underlying conduct, thought, knowledge, and the nature of the Universe. I would disagree with this liberal arts definition and add that a philosophy can also be proven right or wrong by factual observation using the scientific process.

A philosophy enables us to recognize objects and ideas and place them in coherent, understandable categories. A foundation philosophy is one that all other ideas and scientific models are built upon and filtered through. Everything starts with an idea—even the Universe. An idea can develop into a complete philosophy. In the fields of science, a philosophy will create a model that would be tested against known observable phenomena. Philosophers throughout history have dealt with the question of existence. What is existence? What is being? What is reality? Unfortunately, the people who have pondered these great questions were limited by their scope of knowledge and discoveries. The experiences and knowledge we have accumulated throughout a lifetime are the limiting factors that allow us to abstract, conceptualize, or analogize in this reality. These experiences are the tools that enable us to develop a foundation philosophy that will hopefully answer the three most difficult questions in the Universe, which I listed in the Preface. If your philosophy cannot answer these three ultimate questions you better face the obvious—you have nothing, and you better start from scratch. It is better to recognize the fact you do not know then to continue with a dead-end philosophy.

A philosophy is a collection of ideas, which is supposed to create a model to help you understand the Universe around you. If your model of the Universe is incorrect, you have a difficult time understanding the phenomena you see. I give you this analogy to help you to understand the problem. Imagine you have a 10,000 piece puzzle with random shapes and colors, but collectively they form a larger recognizable picture. To make matters more difficult, you do not have a finished picture of what the puzzle looks like when put together. The picture is analogous to a philosophy. If you do not have the finished picture or if you have the wrong picture, then it is much harder to know where the pieces fit together. Without the correct picture, you will need thousands of people working endless hours to piece the puzzle together. The Theory of Multidimensional Reality is a better more accurate picture of how the Universe really works. Some of the conclusions this theory is going to be ego shattering for many of you so put your ego and your degrees aside and think hard about what is presented. The way our Universe was created there is a cyclical time limit for man to understand how the Universe really works, and we are running out of time.

Philosophies come in different forms. Some are religious philosophies, and some are science philosophies. Today these two philosophies are diametrically opposed to one another. The majority of scientists and academicians believe there is no God. The theologians believe God exists in spite of what they may or may not know about Roman history. One would assume that a science philosophy that could explain all known phenomena in the Universe would also prove the existence of God and explain His relationship to His creation. I contend the Theory of Multidimensional Reality accomplishes that.

Science Philosophy

Determinism is the predominant philosophy in current day science. This is the end result of a matter-oriented philosophy of the Universe. It is the pragmatic viewpoint of an engineer who needs to know the cause and effect of his environment so outcomes can be predicted, and products can be built with predictable results. The words science and physics are derived from the word physical that denotes the study of the matter world, the structure of matter and the Universe, including its measurement. The scientific process is about performing experiments and related fact-based theorizing that predicts things that can be verified and repeated. This approach is fine for producing products with consistent performance and results.

There is nothing wrong with this approach to science but it fails when you try to explain the subatomic world or explain the causes of magnetism, gravity, light and mass. What eventually happens is when a scientist discovers a phenomenon that cannot be explained, they just bypass it or come up with a workaround. Many times they just ignore it because its solution would represent too much of their time and money to find the correct answer. The worst situation for the academic status quo is the discovery of a phenomenon that cannot be explained using the "accepted theories," and that to acknowledge such a phenomenon would mean the repudiation of "accepted theories." It could also result in the possible tarnishing of a scientists' reputation and/or the potential elimination of any future government grant money. Besides, the science philosophy they were taught in college gave a plausible answer to many phenomena even if they only worked within a narrow range of instances.

String theories of various approaches have made somewhat of a departure from the quantum mechanics approach to understanding gravity and light. So far experimental cooperation is nonexistent and unprovable. Dr. Jeffrey Harvey of the University of Chicago, who is a proponent of String theory, was reported saying "What am I doing spending my whole career on something that can't be tested experimentally?" I will cover this subject in Chapter Three.

Sir Karl R. Poppers' philosophy on scientific theories was that they can never be proven, only falsified. Therefore, they should be used as calculation tools and not as religious dogma. Admitting something is wrong or "I don't know" is the beginning of a discovery and the end of wasting time and money on a dead-end theory.

There have been over 40 philosophers since the time of the Hebrews and the classical Greeks to the modern age, that have debated the importance between mind and matter. Was mind the dominant force in the Universe or was there nothing more than matter? It appears the atomists had won out. Maybe because we are so attached to our physical bodies we find it difficult to believe there could be anything else in the Universe.

Before I present the thoughts of some past scientist philosophers, I want to give my observation as to whom I would describe as a philosopher-scholar. I have observed that out of 100 rabbis, ministers, physicists, scientists, etc. only about one to two percent are what I would call a real scholar. Most are just practitioners or technicians of their field but do not delve into the "why" and "how" of the most difficult questions. In the modern era there have been a few very notable physicist-philosophers who hinted that the Universe was something other than what they were taught and that there was something more than the traditional four dimensions (length, width, depth and time). Some of these famous scientists knew there was something wrong with the accepted philosophies and wrote about their doubts to friends and in lectures. With that said, the following are without question philosopher-scholars in their fields even though I may not agree with everything they have written and thought.

Sir Arthur Eddington, the outstanding scientist of Great Britain in the Twentieth Century (*The Nature of the Physical World*) stated:

> "Not in the dim past but continuously by conscious mind is the miracle of creation wrought. The idea of a universal Mind or Logos would be, I think, a fairly plausible inference from the present state of science theory. ... To put the conclusions crudely, the stuff of the world is mind-stuff ... consciousness is not sharply defined but fades into sub-consciousness and beyond that we must postulate something indefinite but yet continuous with our mental nature ... it is difficult for the matter of fact physicist to accept the view that the substratum of everything is of mental character."

Albert Einstein wrote:

> "You will hardly find one among the profoundly scientific minds without a peculiar religious feeling of his own . . . his religious feelings takes the form of a rapturous amazement at the harmony of natural law, which reveals an intelligence of such superiority that, compared with it, all the systematic thinking and acting of human beings is an utter insignificant reflection."

The English Scientist, Sir James Jeans, from Cambridge

University, wrote:

> "The Universe begins to look more like a great thought than like a great machine. Mind no longer appears as an accidental intruder into the realm of matter; we are beginning to suspect that we ought rather to hail it as the creator and governor of the realm of matter."

> "we suspect it [the Universe] is all waves and nothing but waves."

> "The old dualism of mind and matter, which was mainly responsible for the supposed hostility, seems likely to disappear, not through matter becoming in any way more shadowy or insubstantial than heretofore, or through minds becoming resolved into a function of the working of matter, but through substantial matter resolving itself into a creation and a manifestation of mind."

Richard Feynman commenting on the state of affairs in quantum mechanics during his November 1964 Cornell Lectures entitled the Character of Physical Law stated:

> "It always bothers me that, according to the laws as we understand them today, it takes a computing machine an infinite number of logical operations to figure out what goes on in no matter how tiny a region of space, and no matter how tiny a region of time. How can all that be going on in that tiny space? Why should it take an infinite amount of logic to figure out what one tiny piece of space/time is going to do? So I have often made the hypothesis that ultimately physics will not require a mathematical statement, that in the end the machinery will be revealed, and the laws will turn out to be simple, like the checker board with all its apparent complexities."

The only scientist-mathematician-cyberneticist that came close to hinting that the Universe was the product of information was David Foster in his 1975 book The Intelligent Universe. He saw the similarity between the DNA and the principles of computer cybernetic control of systems or processes. The important principle he established was the controlling computer that exercises the cybernetic control over something must operate at a much faster processing speed than the process it is controlling. In other words, whatever is controlling our Universe must be operating at processing speeds faster than what it is controlling.

As you can see, some of the most famous mathematician and physicists intuitively knew something was wrong with their matter oriented, scientific theories. I should also include Schrödinger, who also hinted that he did not believe the concept of matter as fundamental. All of these scientists had given up the deterministic mechanical theories of the Universe. Unfortunately, none of them ever took it to the next step by asking themselves "What if the Universe was the product of information? How would we explain what magnetism, gravity, matter, and light are?" I cannot find any evidence where they tried to create any philosophical models that would tackle these weighty questions. They collectively continued to beat this matter theory of existence into the ground.

Today we are faced with the problem of explaining what mass is. For that reason we have two philosophical camps to describe matter and gravity. On one side are the physicists that prescribe to the Standard Model in quantum mechanics and the other camp defend and promote the various String theories, the latest being M-theory. I will later cover these different philosophies, not with the intention of torturing the reader with long explanations, but merely to show that both sides have built a house of cards on talcum power. Both are partially right and partially wrong. I should also note for the reader that both camps had been and are currently dominated by armchair mathematician.

The results of what we know or do not know about our Universe are:

1. Space is curved by large massive celestial bodies.
2. There was a beginning of the Universe we call the Big Bang. We do not know where it came from before the Big Bang, but something cannot come from nothing; there has to be a source.
3. We do not know what gravity is. Einstein analogized it as a form of acceleration in four-dimensional time and space. His formulas are workable but not the complete answer. He did not explain what causes gravity. You cannot know what gravity is unless you know what is mass.
4. Objects get smaller in gravitational fields and time slows down.
5. Whatever matter is, it is made up of waveforms.
6. Planck's constant is the ultimate building block of the energy of the Universe, but no one knows why. It represents one electrical wave.
7. We do not know why the natural log e exists in our Universe.
8. Mass and energy are the same, but we do not know why.
9. We do not know what mass is, which is what causes gravity. Without knowing what mass is, it is premature stating what causes gravity.
10. What is electricity, the electron? Is it a black hole?
11. What is light? Is it a particle or a waveform?
12. What is magnetism?
13. What is time?
14. What is consciousness and life?

You would think that after several hundred years since the birth of the scientific process, someone would have asked the basic question: Something is wrong with our foundation philosophy. Maybe matter is not the dominant 'thing' in the Universe, and there is something else? Only after the last eight years have some physicists come to the conclusion that our reality is a hologram. The traditional theories are currently workable only within a narrow framework. Many scientists have found that their disciplines break down at the limits. Than another scientist comes along to try to expand our understanding based on the earlier work. However, no one has been able to explain the total picture or even to define the most basic definitions of their discipline. We should all try to use a logical approach in science, but the logical approach can only be used where our underlying principles are firmly understood and are correct. Unfortunately, many of the phenomena in our Universe, or reality, are not understood at all. Much of the thought processes and assumptions we employ fail at the limits of our reality.

The following quote mirrors what I am conveying. It comes from the late Imre Lakatos who had a Ph.D. in Philosophy of Mathematics and Science from the University of Cambridge. He stated from his book *Science and Pseudo-Science*:

> "Scientists have thick skins. They do not abandon a theory merely because facts contradict it. They normally either invent some rescue hypothesis to explain what they than call a mere anomaly or, if they cannot explain the anomaly, they ignore it and direct their attention to other problems. Note that scientists talk about anomalies, recalcitrant instances, not refutations. History of science, of course, is full of accounts of how crucial experiments allegedly killed theories. But such accounts are fabricated long after the theory had been abandoned. What really counts are dramatic, unexpected, stunning predictions: a few of them are enough to tilt the balance; where theory lags behind the facts, we are dealing with miserable degenerating research programs. Now, how do scientific revolutions come about? If we have two rival research programs, and one is progressing while the other is degenerating, scientists tend to join the progressive program. This is the rationale of scientific revolutions... Criticism is not a Popperian

quick kill, by refutation. Important criticism is always constructive: there is no refutation without a better theory. Kuhn is wrong in thinking that scientific revolutions are sudden, irrational changes in vision. The history of science refutes both Popper and Kuhn: on close inspection both Popperian crucial experiments and Kuhnian revolutions turn out to be myths: what normally happens is that progressive research programs replace degenerating ones." [Imre Lakatos, *Science and Pseudo-Science,* Pages 96-102 of Godfrey Vesey (editor), *Philosophy in the Open,* Open University Press, Milton Keynes, 1974.]

The Weakness of Mathematics

For the last 130 years, mathematicians have dominated the field of physics. As stated earlier, we still do not know what magnetism, gravity, light and mass/matter are beside the other phenomena listed previously.

Math is an invaluable tool to explain abstract ideas, but it can be misused. In fact, many people who know the power of mathematical description have misused it. However, the real power of math is not in its descriptive application but in our ability to use it to evolve to abstract concepts. This is math's real forté. When you work with numbers long enough they can make anything fit your conceptual view of a problem. This is where the weakness of the "tool" is most pronounced. The final formula is the least important to the math process. The thought process is the most important. If the logic is wrong, the math will not show it. In fact, math can "prove" faulty logic is correct.

The problem with the heavy reliance on mathematical modeling to explain these phenomena is that the mathematician seems not to appreciate fully the fact that higher math is only a tool. The philosophy has to come first. I will explain with this simple analogy. Let us take a building contractor. His tools are saws, hammers, drills, planers, glue, paint brushes, etc. He uses these tools to build a structure based on an architect's plans, but if the architects' plans are flawed the builder will construct a building that may fall or rot away in a few years. The point being that the tools and their correct application cannot show a faulty plan incorrect.

Math is similar. Math is just a tool that can express almost any type of reality the skilled mathematician chooses to use. That is the weakness of the tool. It cannot prove faulty logic incorrect because it can describe almost any kind of reality. An example of this is in General Relativity that concludes the existence of black holes with infinite gravity. The first question that should have been asked is: can black holes even exist in our reality? Another example is with String theory. It is entirely created out of a very complicated long math construct and not from observed facts. With ten or more dimensions and unobtainable experiments to prove it one way or another, it borders on faith, not science. In science, unobserved speculations do not disprove observed facts. An example of such faulty logic is from Dirac, "It is more important to have beauty in one's equations, than to have them fit experiment."[2] The problem with his kind of thinking is that he forgot that the math is merely a tool to understand the experimental results not that the math is the end result.

Where Did the Error Start

The matter oriented theory of the existence is very old in origin. They were first expressed by the Greek atomists, Leucippus (440 B.C.E.) and Democritus (420 B.C.E.). They taught that nothing exists, except atoms and the great void. Plato's book, *The Sophist,* says,

> Some of them [the atomists] drag down everything from heaven and the invisible to earth, actually grasping rocks and trees with their hands; for they lay their hands on all such things and maintain stoutly that that alone exists which can be touched and handled; for they define existence and body, or matter, as identical, and if anyone says that anything else, which has no body, exists, they despise him utterly, and will not listen to any other theory than their own.[3]

This sounds exactly like what goes on in academia today. Non-matter oriented theories of existence existed at the time of Plato and long before him. The Jewish concept of the Universe I would conclude is correct, which is "As long as God thinks the Universe, it exists." Implying the Universe is the product of a thought form, which is correct. The problem the Jews had at that time was that it was difficult finding analogies to help describe their philosophy. Plato used a cave analogy to try to explain existence and the Universe:

> 'And now, I said, let me show in a figure how far our nature is enlightened or unenlightened: — Behold! Human beings living in an underground den, which has a mouth open toward the light and reaching all along the den; here they have been from their childhood and have their legs and necks chained so that they cannot move, and can only see before them, being prevented by the chains from turning round their heads. Above and behind them a fire is blazing at a distance, and between the fire and the prisoners there is a raised way, and you will see, if you look, a low wall built along the way, like the screen which marionette players have in front of them, over which they show the puppets.'
>
> 'I see.'
> 'And do you see, I said, men passing along the wall carrying all sorts of vessels, and statues and figures of animals made of wood and stone and various materials, which appear over the wall? Some of them are talking, others silent.
>
> You have shown me a strange image, and they are strange prisoners.
>
> Like ourselves, I replied; and they see only their own shadows, or the shadows of one another, which the fire throws on the opposite wall of the cave.'[4]

Plato is saying that what we perceive in this reality is analogous to the shadows projected on the wall of the cave.

Man thinks reality is what he sees, but it is really something else.

The philosophies and attitudes of the atomists most closely resemble the philosophies and behavior of modern-day scientists and academia. Solon, Plato, and the Jews fought this kind of philosophy for hundreds of years. Today nobody in academia even questions it. The atomists had won two thousand years ago.

How Our Science Philosophy Was Changed.

I feel it is important people understand why the science was changed after the second century and who did it. Only a very few scholars in Roman history know this information but it is time everyone know why and who.

The Catholic Church taught the philosophies of Aristotle from the second to the eighteenth century. His works were almost elevated to the status of holy. The writings attributed to Aristotle are the origin for the current day matter oriented theory of existence. The academic narrative for the history of Aristotle was that he was a philosophy student of Plato and that he took over Plato's school after his death. In all of Plato's writings Aristotle is not mentioned once. His name is only inserted by footnote by the translators usually stating that Aristotle wrote also this. There was an Aristotle mentioned by other Greek writers but as an admiral in Alexander the Great's navy. The Roman writer/geographer Strabo (his full name is better known as Julius Creaser, Strabo was his family name) also mentions an Aristotle but only as a geographer, most likely the hobby of the Admiral while he was invading other countries with Alexander. Another Roman writer who mentions Aristotle was Plutarch, which was one of the pseudonyms of a very powerful Roman aristocrat. Plutarch repeats the same story written by Strabo. Neither writer mention Aristotle as a philosopher. At this point real scholars of Roman history know exactly who to blame.

The Essence of the Problem

The essence of the problem we face is transitioning from the finite to the infinite. For our finite minds to approach the infinite, we have to use all of our resources. We have been taught that infinity cannot be understood by the finite mind. We have been taught that some concepts are beyond the minds of men. I am going to show you why this is not true, that the perceived complexity of the Universe results from an inaccurate picture of reality. To understand infinity is to understand only what the basic systems are. You cannot understand the basic systems unless you have an understanding of the systems underlying principle.

The best way to conclude this section is a quote from Dr. Lee Smolin, research physicist at the Perimeter Institute in Waterloo, Canada.

> A successful unification of quantum theory and relativity would necessarily be a theory of the Universe as a whole. It would tell us, as Aristotle and Newton did before, what space and time are, what the cosmos is, what things are made of, and what kind of laws those things obey. Such a theory will bring about a radical shift—a revolution—in our understanding of what nature is. It must also have wide repercussions, and will likely bring about, or contribute to, a shift in our understanding of ourselves and our relationship to the rest of the Universe.[5]

I sincerely believe that the Theory of Multidimensional Reality has fulfilled Dr. Smolin's requirements and that it shows that the Universe is the product of information.

Endnotes

1 New York Times article of December 7, 2004. *String Theory, at 20, Explains It All (or Not)* b y Dennis Overbye.

2 Dirac, 'The Evolution of the Physicist's Picture of Nature', *Scientific American*, May 1963, 208, 47.

3 Plato, *The Sophist*, Vol. 7, pp. 371, Loeb Classical Library, Harvard University Press, 1967.

4 Plato, the Republic, Book VII. Loeb Classical Library, Harvard University Press.

5 Quote attributed to Dr. Smolin in 1977 and listed on the Open-Site Web page: http://open-site.org/Science /Physics/Modern/Theory_of_Everything.

Chapter **2**

The Theory of Multidimensional Reality

The Basic Theory

The Theory of Multidimensional Reality is a simple idea, but it just takes time to understand an information theory of existence. In 1973, my friend, Gary Sultan, and I were discussing the idea of the "secret of the Universe," and how long it would take to explain it, and the number of pages it would require. I thought it would be at least a 50 page thesis. My friend thought it could fit on one three by five index card. He was right with room to spare. It distills down to these two simple ideas:

1. Everything in the Universe is the product of information that exists in another time-space relationship that acts like a computer.

2. The Operating System of our Universe is who we call God. He is ultimately the One who creates the information. As long as He thinks the Universe, it exists.

From these two simple ideas, you can build the model that explains how the Universe works. I call it Multidimensional Reality because an object exists in three different dimensions, almost at the same time. The later concept will be explained in Chapter Three.

This chapter covers the basic theory and some compelling proofs. Chapter Three will present how information for an atom manifests itself in the third dimension and becomes an atom. That Chapter will also show what it would look like and behave using the Theory of Multidimensional Reality. From this point on I will abbreviate the philosophy name as MDR.

Creating Analogies to Describe an Information Theory of Existence

The following two analogies will help in understanding the theory. I use two because each provides a particular perspective in describing some aspect of how the Universe works.

A simple everyday television and videotape analogy will define the problem. Imagine you are looking at a television image. Now let us say, for this discussion, that this is a holographic television image that can project a three-dimensional image. Now let us say you are looking at a man, a woman and an outdoor country scene. The images you are viewing do not originate from only within your television. The information that creates the images are stored on a one dimensional-like medium called videotape. The videotape player converts the magnetic information into electromagnetic information. Then it is transmitted from the TV station's antenna to your television receiver. Your television reconstructs the images from electromagnetic information that you see on the screen. The result is the picture of a man, a woman, and a pastoral scene.

Now imagine that you are one of those people in the image. How do you know you are a created being, and your existence comes from somewhere else? How do you know you are in a created reality? Put more bluntly, you are a projection of information from another dimension, but how do you perceive that fact? We face this problem. Now you know why an information theory of existence eluded us for so long. The answer is that you would look for cycles that appear in our reality that logically should not be there.

The Videotape Analogy

If you accept the idea that man is a reflection of the Universe, then our inventions are also reflections of our Universe. Let us take the example of a videotape recorder because this example best describes what is light.

A video camera converts light images from objects into electrical impulses. Electrical impulses can then be stored on magnetic tape or digital storage media, but for this example I will use the older technology of tape. You will notice that the form and dimension in which the picture is stored is different from the electrical impulses that came from the camera—even though the electrical impulses also represent the same picture of the object. Let us take this example a little further. You have a magnetic tape, which contains the information for color pictures with sound. It is stored in a form where time stands still. When the tape is played for broadcast, the information on the tape is again converted into electrical impulses. The impulses go into a transmitter where they change form to become electromagnetic waves transmitted from an antenna. The picture information is now traveling in a two-dimensional form. There is not only the information for the picture but other frequencies as well. There is a center-beat frequency, a carrier-wave frequency (on which the picture frequencies are superimposed), a separate frequency for sound, and finally, a frequency that determines the sweep of your picture screen. The final image you see is a completely synchronized, two-dimensional or three-dimensional "reality." All the analogies for our existence are combined in this one invention. It is ironic that the invention, which is most illustrative of our existence, should be the tool used to mesmerize and brainwash us.

"Here" in one form may not mean "there" in the same form. If we examine a portion of the videotape with an electron microscope, we would see only brownish-red iron oxide. No matter how highly you magnify the tape, we would never see the objects represented by the domains of information on the magnetic tape. If you could identify the objects represented by the domains, the distances be¬tween the information domains, which make up the object, would be micro¬cosmic, compared to the distance relationships in our projected "created reality." The frequencies of the colors, which represent the object, would be at a much lower frequency on the tape, and in a totally different relationship to each other, compared to their image in the projected "created reality." There would be no time perspective for the objects represented as magnetic domains on the tape. They would be frozen in time and space, stopped until passed over by a magnetic tape transducer (head device). The transducer converts magnetic fields of the monolithic structure, iron oxide, into electrical frequencies. The electrical frequencies return to two-dimensional information (the transmission dimension), representing a totally different time-space reality from what was physically represented by the magnetic domains of information. If the tape speed is increased, time will seem to be shorter between two events. If the tape speed is reduced, time will seem longer between the two events. Neglecting synchronization considerations if the tape is increased in speed, the in¬formation reproduced at the receiver (your television set) would appear physically shorter. That is because the receiver's picture-scan is in real-time, and the transmitted information is in a shorter time. Therefore, there will be a shorter object in the picture. This is reminiscent of Einstein's Special Theory of Relativity. The opposite is also true when you slow down the tape speed it would make the object longer. Of course, corrections in the synchronization would have to be made, or the image would appear totally unrecognizable.

If an object is taped with an increased speed with more domains of information on the tape—more mass would be associated with the object. This is the same analogy as increasing the velocity of the object in Einstein's Special Theory of Relativity. So it seems that our representative object not only becomes shorter when its velocity in¬creases but its mass also increases. The Lorenz Transforms seem to describe the same result.

Relationship Between Matter, Energy, Information, and Reality

Using the videotape analogy let us go into how the image is converted into energy and information into our *virtual* matter world. This is all about whether the tape goes through the head device to record or transmit an image. The information-domains passed over by the tape transducer (head device) determine the current reality or existence. The velocity or speed of the tape determines time-space relationships and the rate of which information is transmitted.

The Special Theory of Relativity was called special be¬cause he recognized the speed of light (2.997925×10^8 meters per second or 186,000 miles per second) as the constant of the Universe, and that nothing can go faster than the speed of light. There were other relativity theories at the time, so the word "special" identified Einstein's. He theorized that no matter how fast an object was going, if a beam of light were sent out from the object, the velocity of the light would be the same as for an object at rest. This idea was a departure from classical mechanics, which taught the cumulative effect of velocities.

The mechanism that contains and transmits our Universe's information is not like a videotape. It is a very special kind of computer that I call the Diehold. The videotape analogy does help explain what light is and why it is the constant of the Universe. The speed of light is analogous to the speed of the magnetic tape across the head device. Using the computer model, it would be the result of the input-output (IO) speed of the computer.

What is Light and What is Time?

Light is the result of the collective spectral lines (information) that make up the source of the light. The spectral lines, quantum mechanics and string theory will be covered in Chapter Three. The matter world we live in is the object being propagated into existence at what we call the speed of light. That means the light (photons) are really "motionless" information on the "tape" receding from us. We perceive it as it recedes from us. Our world and the Universe is the object that is being created at what we call the speed of light. Time is then the rate at which the Diehold transmits our information. It is not another dimension but rather a function of the Diehold.

A simple analogy: Imagine you are in a plane flying at 400 mph. You are looking out the back of the plane while it is releasing a contrail behind it. To your point of view, the contrail cloud is traveling away from you at 400 mph, and you look as though you are standing still. The vantage point of someone on the ground shows the contrail speed to be zero, and your plane is flying at 400 mph. Light is like the contrail falling behind us. The only difference is that in our dimension the light, or information, leaves this dimension 360° around the object giving off the light or reflecting the light. The light is in little packets reflecting the pulsing of the frequencies creating the atoms. I will cover this in greater depth in the next chapter when I cover Planck's constant.

Now I do not want you to think that I am saying we are all on videotape, and everything has been set in time. That is not the case. The Diehold is a computer, so rather than a "head device" there is something that is acting like a Central Processing Unit (CPU). There must also be something that functions like our video cards, which is to combine various types of data and combine them to a three-dimensional universe at a slower rate than the CPU. There would be a delay in the transmission of the data at this stage. This idea will be expanded upon later.

The Computer Analogy

The computer analogy reveals the greatest secret there is in geophysics, astronomy, and climatology. This discovery reveals what causes the ice ages, geomagnetic (polar) reversals, mass extinctions and the creation of new species and why they all happen cyclically through geologic time. I had been studying the causes of the geomagnetic reversals since 1970. I knew they happened roughly every 12,000 years. I had covered detailed research results to these questions in my 2007 book God's Day of Judgment the Real Cause of Global Warming (Chapter 8). I will briefly use some of those discoveries to show how this Diehold works. My studies included reading all the journals Science, Nature and Geology from the mid-1950s to the 1990s. The textbooks were also useful, but I found many errors in them when compared to the discoveries published in the journals. Somehow published discoveries that countered what was taught in the textbooks never make it into the new textbooks.

The Main Clock Cycle

The Diehold contains and creates all the information of our entire Universe. It creates the matter world we occupy. In essence, our Universe is a created reality. It is similar to a hologram, except this hologram creates the matter world we live in, at least what we call matter. Like all computers, the Diehold would have to have clocking, synchronizing and resynchronizing frequencies. Clocking frequencies are stable oscillator pulses in a synchronous computer used for timing of all operations: such as gating, recording, input/output functions, floating-point operations, video, etc. The bits of information in a synchronous, or sequential, computer will wait until a clocking pulse arrives, then the bits of information will proceed to another section of the core memory, or input/output functions, or to the software in RAM memory. Computers produce these frequencies to keep massive amounts of bits of information within specific time slots. All of these various synchronizing frequencies times to the main clock cycle. This is how all digital computers work, and there are no exceptions. Computers must have these synchronizing and resynchronizing frequencies to prevent a race condition, which is when the computer gains or losses bits within a clock cycle. A computer could not work unless it had these clocking and synchronizing frequencies.

The Diehold's main clock cycle and synchronizing frequencies manifest themselves in this dimension in several ways. I discovered the main clock cycle in 1989. By 1988, I had compiled a database of all known stars, open clusters, planetary nebulae and globular clusters (Appendix A). I compiled the list from two well-known sources: Sky Atlas 2000 by Sky Publishing and The Cosmological Distance Ladder: Distance and Time in the Universe, published by W. Freeman and Company. I carefully analyzed the distances in light years that were close to these timelines and found some discrepancies, but I am confident that the list is correct. It includes over 520 entries, representing over 8,675,000 estimated stars, ranging in distances from 65 to 304,000 light-years away. Graph 2-1 clearly shows six "blank periods" in light years where no stars are visible. Table 2-1 shows the numerical results of the database, from 11,089 light years to the end. Notice the six "blank periods," almost all spaced 12,068 light years apart, and they are about 1,000 years in duration where no stars are visible. There are stars in the blank periods, but the reason we cannot see them is because they all novaed on the 12,068-year clock cycle, and the dust conceals their location.[1] This graph shows that all stars nova at the same moment in time, all over the observable Universe. The only way this could occur is if the entire Universe was the product of information, and the main clock cycle affected everything, everywhere at the same time.

Reversal No.	Distance in LY	Distance of No. Stars Visible (LY)	Distance from last blank period
1	11,742	1,304	N/A
2	23,810	1,630	12,068
3	35,878	977	12,068
4	47,945	978	12,067
5	59,360	1,304	11,415
6	71,428	5,870	12,068

Table 2-1: The table illustrates six blank periods as seen from the earth, where no stars are visible for various distances.

The astronomical distances are determined by the red shift of the star's light. I asked an astronomer how much error appeared in the formula that calculated distance. He said the error factor could be as much as 10 percent. Distances are compared to a known star that is relatively close to Earth. I wanted to see how accurate the first distance of 11,742 light years was. If there was an error it would show up there, because we are measuring from a fixed position to a variable position. The relative positions between the other blank periods are correct because the error is incorporated in both distance calculations. It turns out that the error was only 2.2 percent, so the blank period started about 250 light years further out.

It is impossible to have any blank areas around us, not to mention that four of the six are each about 12,068 light years apart! So what causes these blank periods?

At the time of the polar reversal, all stars nova at the exact same time, on the clock cycle and that is what causes

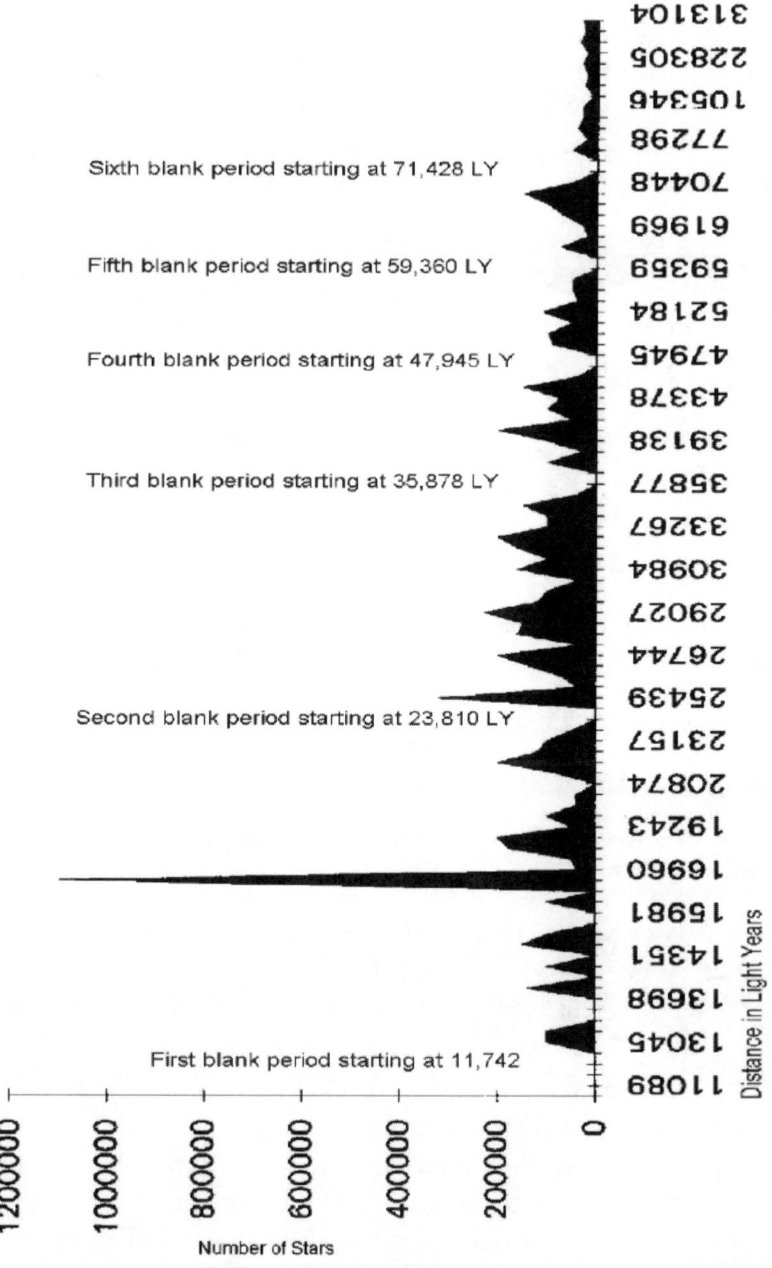

Graph 2-1: The Graph clearly shows the six blank periods I discovered in our galaxy. The x-axis is in light years from Earth and the y-axis is the estimated number of stars cataloged at that distance.

these blank spaces in time. When a star novas, it throws off its surrounding matter shell into space at a high velocity. I came to this conclusion after reading the discovery of Bok Globules (Figure 2-1). The astronomer, Bok, had discovered over 450 of them, in all directions from our solar system. Their distances ranged between 13,000 to 14,500 light-years from Earth.

Figure 2-1: The dark spots are the Bok Globules. They are found in all directions from the Sun.

In later years, some of these stars had been classified as planetary nebulae. Figure 2-2 shows some of these planetary nebulae. You can clearly see the dust cloud around the star in the center.

The traditional explanation for Bok Globules is: They are young stars just forming from the intergalactic dust surrounding them. The problem with this theory is that it is contrary to the basic idea that galaxies start as quasars and then shoot off two jets of stars which eventually form the spiral arms of the galaxy. That means the oldest stars are located on the outside-edge of a galaxy, the younger stars near the center. Our Sun is located about 30,000 light years from the center of the Milky Way therefore it is fairly old. It does not make any sense to have

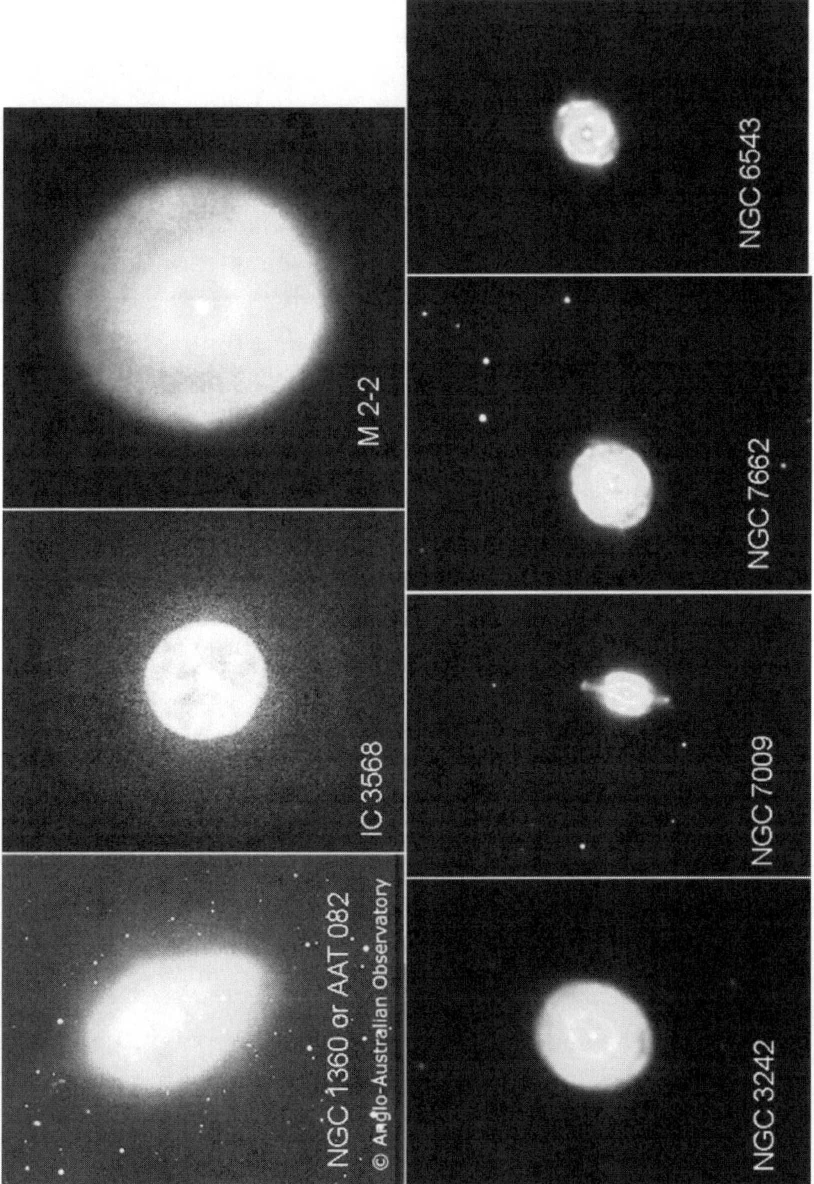

Figure 2-2: Picture of seven stars some time after their novas.

very young stars in the center, and also further out with very old stars. Bok Globules are found in all directions from our Sun. The discovery of star V838 in 2002 settled the issue because it clearly showed that we are looking at a star at various stages after it novaed on one of the time lines 24,000 years ago.

The traditional explanation for a stellar nova is when a star novas it is the end of its "life" and would become a neutron star or black hole. The evidence is obviously contrary to this theory, but academia still sticks to their incorrect philosophy. Our own star's nebula was discovered in 1984 and later confirmed by NASA in 1989 when the Voyager-2 photographed some of the remaining dust rings. The name given to this remnant dust ring is the Kuiper Belt and is located from Neptune to Pluto. Since then nine other nearby stars have been discovered with similar Kuiper-like belts around them.

Next I wondered that when a star novas, is it possible that it only produces 500 to 2,000 times more energy output in one day. With the discovery of Bok Globules, I realized that when a star novas the matter shell around the star is expelled and obscures the light of the star for about 1,000 years. When the dust shell becomes thin enough the star becomes visible once again, as shown in Figures 2-2 and 2-3. So what I had to look for were blank places in space, where no stars were visible.

Cooperating Evidence

I have been studying the causes of the ice ages and geomagnetic reversals since 1972. Those results were included in my 2007 book. In that book, I showed six ice ages occurring after geomagnetic reversals. There was ample C^{14} data from glacial till that indicated a 12,000 period between successive ice ages.[2]

The Synchronizing and Resynchronizing Frequencies

It was easy and surprising finding the synchronizing and resynchronizing frequencies so fast. I did not have to look further than our Sun. When I wrote my first book, I knew the sunspots were the result of a resynchronizing frequency in the Diehold. I also assumed they would all be timed to the main clock cycle. While doing an unrelated research project, I came across a longer solar cycle called a Gleissberg cycle. The Gleissberg cycle consists of eight sunspot cycles (Table 2-2). The first sunspot cycle in the series has the lowest output the eighth has the greatest. The sunspot cycles are

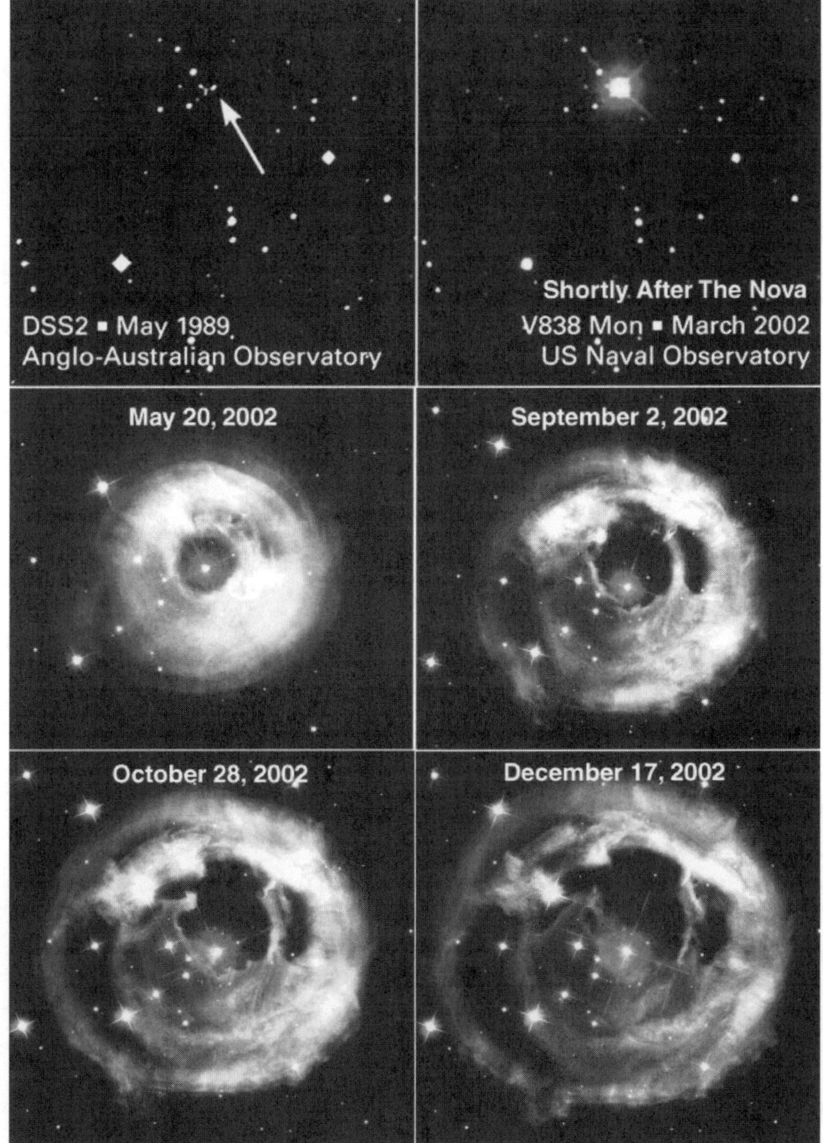

Figure 2-3: Nova photographs of star V838 from February 2002 to December 2002, over an 11 month period.

the resynchronizing frequencies, and the Gleissberg is the synchronizing frequency. The last Gleissberg maximum was solar cycle number 19 occurred December 22, 1957, to March 31, 1958.[3] The Gleissberg cycle turned out to be 88.735 years.

So I divided the 88.73 years into 12,068 to see if the result would be a round number, and it was—136 Gleissberg cycles totals 12,068 years (88.7357 years × 136 cycles=12,068 years). What is amazing about the number 136 is that it is very close to the current value of 137 for the Fine Structure Constant.

I really do not need any more proof that the Universe is the product of information with the discovery of the main clock cycle of 12,068 years and connecting the sunspot cycles to the same number of years. We are living in a created reality. We are looking at the business end of the clock cycle and what had happened to man and Earth in the distant past. I know that old ideas die hard, and egos are threatened by change, but science will have to change just as they did with the failed theories of phlogiston and the aether, both turned out not to exist.

Redefining Dimensions

Multidimensional Reality defines dimensions much differently than traditional physics teaches. Quantum mechanics teaches there are four dimensions (length, width and depth) plus time as the fourth. String Theory (M-Theory) says there are ten dimensions plus one for time. The Theory of Multidimensional Reality concludes that there are a total of eight dimensions and time is a function of all of them, so it is not a separate dimension. Time is the rate at which the Diehold transmits the information for seven of the eight dimensions.

The First Dimension

The first dimension would be the bits of information in the Diehold that contains our information. Time, in essence, stands still for the bits of information. I do not classify time as a separate dimension but rather a function of all the dimensions. Time distils down to the rate at which the Diehold transmits all information because time is a function of the speed of the part of the Diehold that assembles our reality. What I mean by this is the Diehold must have a section in it that functions similar to our video boards, also known as a graphics card, in our computers.

A graphics card receives various types of digital information from the CPU (images, sound, color, dimensional information). It converts it to pixels, sound, and sends it to the monitor through a cable at a slower speed than the operating speed of the CPU. The Diehold must also have a part of it that serves the same as our video boards. Its function would be to combine a great deal of information and assemble it to create our reality. The reason is because the CPU has to function at speeds of the atoms which appears to be Planck's Time expressed as:

$$t_{\mathrm{P}} \equiv \sqrt{\frac{\hbar G}{c^5}} \approx 5.39106(32) \times 10^{-44}\ \mathrm{s}$$

$\hbar =$ the reduced Planck's constant; $G =$ gravitational constant; $c =$ speed of light; s = seconds.

If everything was created and "lived" a lifetime at these speeds, we would live an entire lifespan in less than one second. So I have to conclude that there must be something functioning like our graphics boards which assembles our reality at a much slower speed for living creatures. This time difference is important and will be covered in the next chapter when I cover ex the natural logarithm.

The atoms are the building blocks of our reality but living conscious entities are the main reason the Universe was created. Both the atoms and the information that makes up the conscious entities reside in the Diehold. The first dimension must also have RAM (random access memory) that stores all events in time and also a memory controller that addresses all this information. I consider all of this the first dimension. It is irrelevant for this discussion if the information is stored in a crystalline structure or as bits on transistors like our modern day RAM. It would be safe to say it is some sort of quantum computer.

The Second Dimension, the Transmission Dimension

The information from the Diehold is then transmitted into the second dimension, which is the transmission dimension as explained earlier. The following Graph 2-2 illustrates what

I theorize. The *y*-axis represents the transmitted information in the second dimension originating from the Diehold. This waveform represents the carrier wave that some or all the information is strapped onto. This graph and resulting model will be covered extensively in Chapter Three.

The Third Dimension of Matter

The third dimension is when the information becomes an atom of matter like a carbon atom. This is the dimension of length, width and depth. The z-plane shown in Graph 2-2 is our matter world. When I first developed this graph, I almost immediately realized that those spikes going to the right or left (plus or minus) are our matter world. It is just like Planck theorized, and I realized that what I was looking at is a graphic representation of Planck's constant defined as discrete packets (quantum's) of energy given off by atoms. The only major difference between the Quantum theory and my MDR theory is that these spikes or quanta create our matter world and what we call matter. Our matter world is energy contained

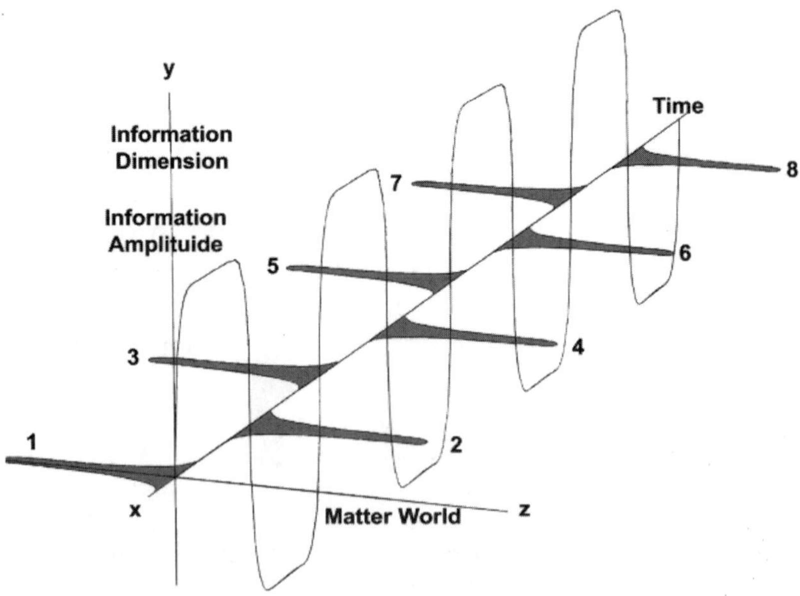

Graph 2-2: The second dimension (y), the transmission dimension creates the third dimension (z).

by information and what we call mass is formed information of energy as it is directed to a point in time and space. Like the light, it is detected as packets of energy as it leaves our dimension.

The Fourth Dimension

From this point on the different dimensions are delineated by the amount of information and potential (voltage) an entity can access, possess and control. Potential should be considered synonymous with information in the Diehold.

The Fourth Dimension is the world of living things. A consciousness is "married" to a physical body made up of atoms. A more technical way of describing the process is the physical body is transmitted to the same coordinates as the information that makes up what we call the soul. This process would include everything from one-celled animals to humans. The primary quality is that these life forms must physically touch matter to manipulate it and interact with it. They must consume food to maintain their bodies. Their senses are how they communicate with others of their kind. For mankind, we must use our five senses of smell, taste, sight, hearing/sound, touch to interact with our world. Man is a fourth-dimensional being.

The Fifth Dimension

The Fifth Dimension would be an intelligent being that possesses much more potential than the previous dimension and can perceive, control and manipulate the world around them without the need to physically touch the matter world. Using their consciousness in the Diehold they could manipulate matter the way they want. That means moving it without touching it by merely changing the coordinates of the information that makes up the matter they wish to move or alter. They would be able to communicate with others by thought only, without physically creating sound. They would communicate directly with another fifth-dimensional individual using only their minds through the Diehold. This type of being would perform much of their interaction (information gathering) through the

Diehold, without the necessity of their physical senses. They would still have a physical body, and they would still have to care for and maintain it, but the mind-body connection would be much different from fourth-dimensional beings. Fifth-dimensional beings could actually "will" themselves well and perhaps live for hundreds of years. After all the only reason, we grow old is because the end caps on our DNA wear down and do not reproduce correctly. If their minds can prevent that I could see where such an individual could live for many hundreds or even thousands of years.

The Sixth Dimension

The Sixth Dimension being may not need or have a physical body at all, but may be able to create one if it pleases them. They would just be pure information in the Diehold. They would have the ability to move objects with their thoughts. Time for them may not be relevant to them because they can "live" forever because, after all, they are a process in the Diehold and no physical body to deteriorate. They would be able to remember everything they ever did no matter how many millions of years it was previously. A sixth-dimensional being would possess greater potential than the previous two dimensions. They may have the capacity to create three-dimensional objects, and take the form of some fourth-dimensional life forms. Sixth-Dimensional beings probably perceive all of their surroundings from the Diehold, and not from our dimension. This does not mean they do not need third and fourth-dimensional objects to survive. They probably use them as reference points, so they know what their information is doing in the Diehold.

The Seventh Dimension

The Seventh Dimension are planets. Since they occupying time and space, you must be able to explain their existence. They possess huge amounts of energy potential, as demonstrated by multiple frequencies emanating from them. Planets interact with their space-time surroundings, existing for great periods of time. They most likely cannot perceive the building blocks of our matter world because they cannot perceive extremely short periods of time. The theory of MDR concludes the stars create the planets. I have also discovered that they share some of the same internal frequencies. I discovered this by comparing the number of sunspots to an increase in earthquakes on the Earth. What I found was that during sunspot maximum when there was an increase in sunspots or eruptions on the Sun there was an increase of earthquakes 240± days after.

The theory of MDR states the heat and frequency sources of a planet originate at the center where there is a center modulation point where massive amounts of information that makes up the planet is directed. That means the center of the Earth would have something that looks like a very bright light source, almost like a star. The matter for the earth would not start forming until it reached the outer core because the information/ energy from the center modulation point would still be too high to permit matter to be formed. The reason I came to this conclusion is from earthquake data. The P-waves travel faster through the center of the Earth than the S-waves that propagate in the crust. I contend that this center modulation point creates gas plasma under extreme pressure and that such a gas plasma would conduct the P-wave motions as good as a solid iron core, which is what is currently theorized. The second reason is that the P-waves are picking up other frequencies after it passes through the center. This process is called heterodyning which is when you have one frequency, and you mix it with another, you get a third frequency as a result. The Earth's magnetic core is created by how the information is phased into our existence and it reverses at the time of the geomagnetic reversal every 12,068 years. The problem with the current solid iron magnetic core theory is that iron loses its magnetic properties at the Curie temperature of 1060°F. They assume that the Earth's core has a cooler temperature than the outer core, and mantle that just does not make any sense since the heat would transmit throughout the entire core.

The last proof I will present is that satellites have encountered radio noise as they cross within five degrees north or south of the equator. The radio noise is formed 90° out of phase to the direction of the magnetic field of the Earth. This could only be formed because there is an electrostatic source located in the center of the Earth and not from a solid iron core.

The Eighth Dimension

The Eight Dimension is stars. The stars create planets, and they possess vastly greater energy than planets. Stars exist much longer than planets and like planets, they cannot perceive the building blocks of our matter world because they cannot perceive extremely short periods of time. We know the stars exhibit a myriad of frequencies that cannot be explained using nuclear fusion or fission.[4] The theory of MDR states the heat and frequency sources of a star or quasar originate at the center where there is a center modulation point. All the massive amounts of information that make up the star are directed to the center. This is the same mechanism for planets in the seventh dimension including the presence of solar radio noise. One of the differences stars have it they exhibit a ring of planets along its equatorial plane.

The quasars would not be considered an eight dimensional object because it is more like a portal opening where the Diehold pours out information to create the stars. Such a portal to the Diehold would exhibit energy levels greater than could be explained with our current physics models.

Gravity

Traditional Quantum Mechanics cannot explain what mass is, which of course brings us to what is gravity? One has to have a mass in order to have gravity but in Quantum Mechanics when they try to accommodate for mass by using a field their results go to either a positive or negative infinity which makes no sense to them. To me it makes a lot of sense because the math equations are telling us when it goes to infinity it is going to or from another dimension 90 degree out of phase from our reality as the graph shows. String theory was developed because it supposedly held out the hope that physics could explain gravity by using either 10 or 11 dimensions and some other difficult assumptions that could never be tested. To this day String theory, remains a very difficult mathematical model that can never be proved by empirical experimentation.

So what is gravity? Is it a pulling action or a pushing action? The General Theory of Relativity depicts it as a pulling force toward an object. The answer using the theory of Multidimensional Reality is that it is a pushing action caused by the information that make up a planet or star being directed to a point in time and space. The information that makes up everything in its vicinity will be "pushed" by the larger bodies' information towards its point in time and space. The need for assuming the existence of neutrinos or Higgs particles or Higgs fields is unnecessary because what we call gravity is all happening in the first and second dimension.

The traditional Einstein explanation of gravity is the result of curvilinear acceleration in three dimensional time and space. Gravity is viewed more as a product of momentum and inertia in space. The General Theory of Relativity predicts that objects would decrease in size and time would slow down, as objects approached a large gravitational object. Time and size changes could only be measurable near objects having super gravities—when they are near a black hole, for example.

I decided to test my theory of gravity against the General Theory of Relativity. I chose several gravitational anomalies located in Oregon and California to investigate.

I have included some of the results in two of my earlier books on the subject.[5] The results of my experiments are these: A six-foot pole shrank 5.8% over a distance of only seven feet. Time changed dynamically inside the vortex, both slowing down and speeding up. It took greater force to push a dead weight towards the center of the vortex, than away from it. The operators of both of these vortexes told me that very few science professors ever visited their sites, but the ones who did, told them there must be a large iron meteor buried deep down in the ground. The problem with this theory is there is no evidence of an ancient meteor crater or any large deposits of iron in both areas. There are two primary reasons their explanation of gravity is wrong. I discovered from my experiments that time and size changed dynamically. The only way you can explain this employing the traditional General Theory of Relativity, is if the large dense body, supposedly buried there, would have to be dynamically changing its mass, which is impossible. For my time measurements I used a 25 MHz clock standard and a frequency counter located outside the vortex. What is revealing to me is that the Santa Cruz vortex is very close to both Stanford and Berkeley Universities, but none of their physicists were curious enough to study the Santa Cruz Mystery site.

Gravity Wave Experiments that May Prove the Universe is a Hologram.

Einstein's General Theory of Relativity predicted gravity waves in the Universe. Several experiments were setup to test this theory. The Germans gravitational wave detector observatory, called GEO600, started in 1995 looking for gravity waves (frequency range of 50 Hz to 1.5 kHz.) The detector consisted of a laser beam reflected over 1200 meters. Since its inception it has not detected any gravity waves, but it did detect "noise" that could not be explained until the idea was proposed by Juan Maldacena using M-Theory with 10 dimensions. The "noise" was the result of the possibility that our Universe was a hologram. U.S. Department of Energy's Fermi National Accelerator Laboratory (Fermilab) Director Craig Hogan was so impressed by GEO600s findings that Fermilab upgraded their gravity wave detector to replicate the same experiment. The new equipment is called a Holometer and will be testing for frequencies in the millions of cycles per second.

By 2003 many other physicists had concluded that our Universe is a hologram so it is not a strange or far-out theory thanks to string theory. I had concluded it and described how it would work in my first book Reality Revealed (1978).

Quasars and where the energy comes from.

Quasars are believed to signify the birth of a galaxy. Energy levels from quasars are greater than can be explained by traditional physics. Quasars emit huge amounts of energy (10^{58} ergs), which is the equivalent of consuming 500 million of our suns per second to produce the energies emitted throughout the electromagnetic spectrum.[6] A good summary of the problem traditional astrophysics face is found in the following quote from the astronomer, Dr. George Abell:

"We have than the perplexing picture of a quasar: an extremely luminous object of small size displaying enormous changes in energy output over intervals of months or less from regions less than a few light months across; 100 times the luminosity of our entire galaxy is released from a volume more than 10^{17} times smaller than the galaxy."[7]

This contradiction between the philosophy that energy comes from matter, as exemplified by the famous equation $E = mc^2$, compared to actual observation on the cosmological level, appears to have a fatal problem. If our galaxy, like all galaxies, went through the quasar stage of evolution some 15 billion years ago and had consumed matter-for-energy to produce the power quantities mentioned before, the Milky Way in its quasar stage would have used up all of its lighter atoms (certainly all of its hydrogen and helium) billions of years ago. To put it simply, this galaxy and our sun should not exist today.

This discovery happened between 1963 to the mid-1970s, and it was a major problem for physicists. The current science philosophy falls apart when it researches its limits such as in this case for astrophysicists. The discovery of quasars was their limit of observation, and they could not explain what they were seeing. Some physics and mathematicians thought that there was a black hole in the center of the quasar, and the energy was the result of matter entering the event horizon and producing that energy. The scientist that came up with that idea was John Wheeler (who also coined the term black holes). Later scientists and mathematicians Stephen Hawking, Roger Penrose, and Dennis Sciama, also worked on the problem of black holes and quasars. In 2014, Steven Hawking changed his position and now says black holes do not exist as we thought of them before.

The idea that a black hole could even exist originated from

Einstein's General Theory of Relativity even though Einstein did not believe one could squeeze atoms that tightly together to create a black hole. In 1939, Robert Oppenheimer suggested that a star could collapse down and create a black hole. The first thing that should have been asked is: Can we even have a black hole in our reality? It looks like Einstein, and I say no.

The problem with the traditional explanation for the energy source of quasars and the center of galaxies is that it goes counter to what we were taught in college. Oppenheimer's idea that at the end of a star's life it would collapse down to a neutron star or black hole if the star's mass was great enough. The idea that a black hole could be at the center of a galaxy is contrary to this theory. Quasars are the beginning of a galaxy, and we see two jets of stars emitted from them (Figure 2-4). No one has yet

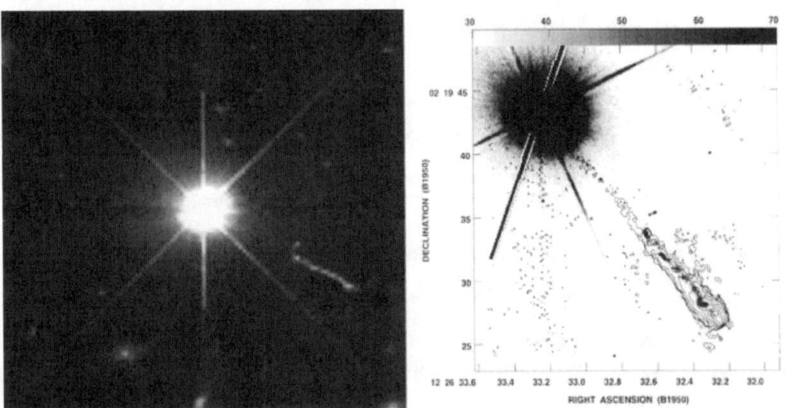

Figure 2-4: Quasar C273 is showing one of the jets of stars coming from the center.

explained how two jets of stars could be coming from a black hole from the center of a quasar or galaxy. It is obvious that the physics community was looking for any excuse to explain the energy output from quasars.

Maybe the Diehold, when it creates a galaxy, opens up a portal and rushes a tremendous amount of information to that point in time and space thereby appearing to emit more energy than can normally be produced in this dimension.

Kirilian Photography and the Phantom Leaf Effect

Not too often can an experimental piece of equipment,

costing less than $100, prove a theory with one phenomenon. The phantom leaf is a phenomenon caused by passing high voltage (low current) at a high frequency (ranging from 3,000 hertz to 5 MHz) through a leaf with a photographic plate on one side. It is not really photography using cameras or lenses. It is a high-voltage, high-frequency effect. The electron discharge is what exposes the photographic film. Figure 2-5 shows a cross-section of how a unit is built.

The phantom leaf effect is produced when 2% to 10% of a leaf is cut away and the remaining part placed between the plates within a very few seconds after the cutting and then photographed. The uncut leaf never touches the photographic film before the part is cut away. The phantom images are not a residue from the leaf before it was cut off. The first time it ever comes in contact with the film is after it was cut. The leaf was immediately photographed and the phantom may appear if the correct frequency is used. Some scientists have tried to explain away the phantom leaf effect by claiming that the whole leaf must have come in contact with the photographic film before the picture was taken. While untrue it demonstrates how scared some scientists are of this phenomenon because they know what it means. When the leaf is properly "photographed," with the correct frequency, the cutaway section of the leaf will appear as if it was still there. It only occurs about five percent of the time, and that is because we do not know all the variables involved

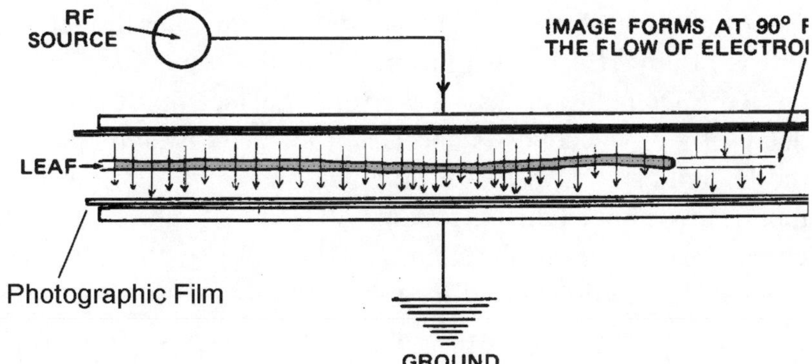

Figure 2-5: Cross-section of a Kirilian "photographic" unit, where the leaf would be placed.

when setting the frequency source to produce the image. The photograph must be taken within a very few seconds of cutting the leaf. Figure 2-6 and 2-7 are both photos of two leaves, where part of the right side had been cut off.

I have indicated where the cuts are in both pictures. The surface of the leaf looks similar to small bubbles and streamers of light coming from the remaining part of each leaf, even from the areas that no longer physically exist! On the missing portion, the resulting light pattern outlines the superstructure of the leaf, all the way to the edge. It is not often that this type of phantom appears. Figure 2-8 shows a photo where the top of the leaf was cut off. Only streams of light outlining the former leaf shape are visible.

Some experimenters have reported a 5% success rate in "photographing" phantoms. The phantom image is very amazing to observe in a real-time environment. When photographed with a video camera, it is seen to fade in and fade out of the picture for up to five to eight seconds before it disappears altogether![8] The phantom leaf effect is a repeatable phenomenon.

Multidimensional Reality Explanation

First, you must ask yourself, "How does a leaf know how to recreate its original image?" If you consider what the leaf has going for it, in this dimension, the answer is no. The only possible answer for the appearance of the phantom effect is that the information for the whole leaf exists in another dimension and is briefly still being transmitted to the same coordinates until the Diehold corrects the error. The fact that there is a delay of a few seconds is another proof there is some part of the Diehold that acts like a graphics board that operate at a much slower speed than the main CPU. The life and conscious energy part of its information, (its "soul"), exists in another time-space relationship. Since the only way to make this phantom appear is by using a high-voltage, high-frequency device, this implies that the signal that makes up the leaf is a modulated signal of some sort, otherwise we could not possibly make the image appear. As we look at the phantom portion of the leaf, we notice that the bubbles of light and the aura around the edges are quite visible. The only thing that is not present is the light that makes up the actual matter in this dimension. That

Figure 2-6: A Kirilian photograph of a cut leaf that exhibits the Phantom Leaf effect. Courtesy of T. Moss, UCLA

means we are looking at the portion of the leaf's information that makes up its *conscious energy*. Another way of stating it is, the information that makes it a living entity. The pulsations observed from the phantom are the result of these different frequencies dynamically changing.

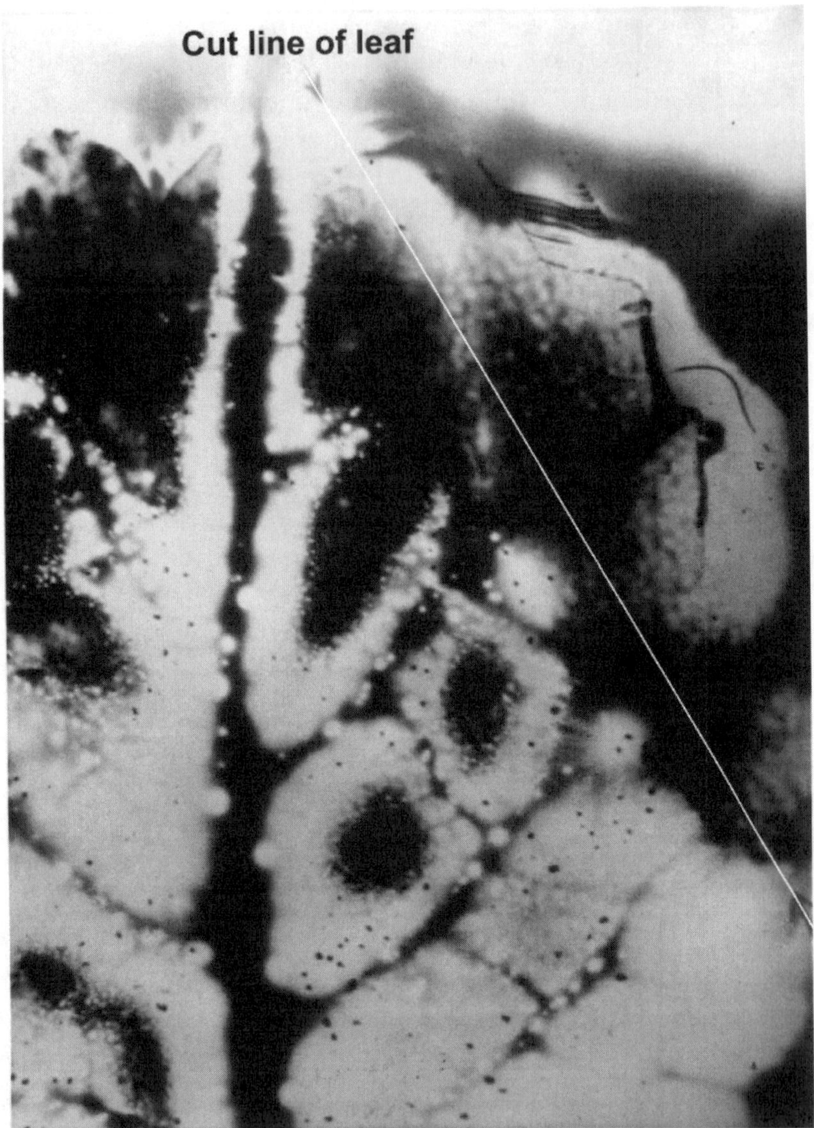

Figure 2-7: A Kirilian photograph of a cut leaf that exhibits the "Phantom Leaf" effect. Courtesy of T. Moss, UCLA

The phantom leaf effect proves two important principals: first, that energy in the form of photons does not always come from the change in the energy level of the electron shell around an atom. In all three pictures, we have light coming from the absence of matter. So does that mean $E = mc^2$ is still valid in

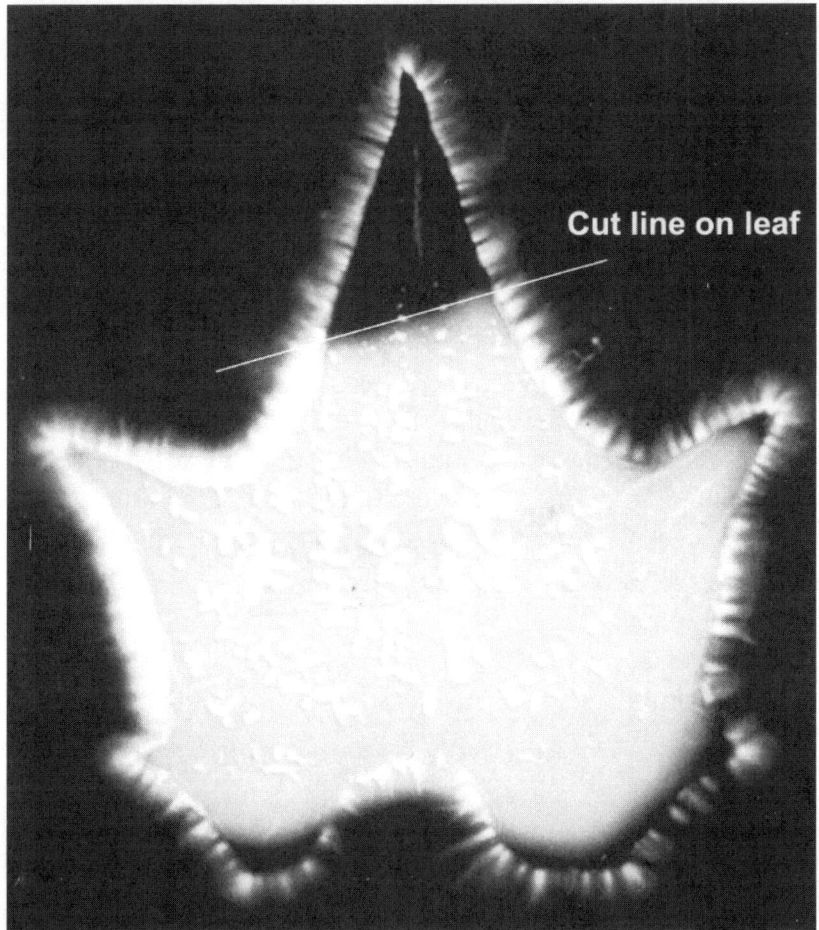

Figure 2-8: Phantom leaf effect from a cut leaf courtesy of Hernani Andrade, Brazilian Institute of Psychobiophysical Investigation.

this case? The second point is that our consciousness exists in another time-space relationship. The information that creates the consciousness of the living object still exists in the first dimension. It was still being transmitted to the same coordinates for a few seconds before the Diehold corrected the transmission.

The important conclusion of this simple experiment is that when we die our consciousness still exists in the Diehold but we no longer have a physical body until our soul is reincarnated another time.

Proof of the Carrier Wave of Existence

Earlier, I theorized several times that there is a carrier wave the information is strapped onto. This is the hardest phenomenon in the Universe to discover and figure out but ironically, it was the first one I figured out, and it is the one that led me to write my first book, *Reality Revealed, the Theory of Multidimensional Reality*. The question is: How do you discover the carrier wave if it exists?

Mysteries of the Pyramid

During 1973 to 1975, I was experimenting with pyramids and studying them for their effects on living things. I am not describing any shape of pyramid, with any type of angle. I am talking about pyramids that were built in the same proportions and angles as the Great Pyramid of Giza. This formula creates the correct angle:

$$\text{Pyramid angle} = \text{arc cos}\left(\frac{1}{\sqrt{e}}\right) = 52.66093239°$$

The following are the unusual phenomena associated with a pyramid built with the slope angle:

- The pyramid can be built out of any metal or stone material and still and have an effect on whatever you place inside it.
- Food kept under the pyramid will stay fresh for two-to-three times longer than food not in the pyramid.
- Artificial flavorings in food will lose their taste, but natural flavors are enhanced.
- The tastes of foods change; they become less bitter and acidic. When we take a spectrographic reading of the treated item, it shows a change in the molecular structure.
- The pyramid will dehydrate and mummify organic objects, but it will not permit decay or mold to grow.
- There is a slowing or complete stopping, of the growth of microorganisms.

- The Swedish Scientist, Dr. Carl Benedicks, discovered that the pyramid produced a resonance or frequency inside. Two German scientists, Born and Lertes, also discovered that this frequency was in the microwave range.[9]
- Researchers say that items placed under a pyramid stay "charged" for various lengths of time after being removed from the pyramid. It has been found that water keeps its "charge" longer than any other molecule, and water will lose its chlorine taste and generally taste better.[10]
- Plants grown under a pyramid grow about twice as fast, in their early life, than plants that are not grown under a pyramid. The treated young plants look healthier and have less insect damage.
- It has been found that a copper pyramid has the best effects and intensifies the effect on organic materials.
- The pyramid has an effect on inorganic items. It is well known that razor blades keep sharper longer, if placed under a pyramid between uses.
- Bill Kerell[11] has done many experiments using brine shrimp, which usually live six to seven weeks, but under a pyramid, Bill has kept them alive for over a year. He also noticed that pyramid-grown shrimp grew two-to-three times larger than normal.
- It has been found that the theta and alpha brain waves are increased. These frequencies are also higher and the signal strength is twice the normal amplitude.
- Inside a pyramid, a tingling sensation on the skin, similar to that of mild electricity occur with an increase in the skin temperature. There is a tranquilizing effect on the nervous system, a deeper "dropping off" in the transcendental state and finally, very graphic dreams occur in vivid color.
- This discovery I know personally. A woman's menstrual cycle went from 30 days to 45 days.

The Explanation of These Mysteries

The general explanation is this: You are in essence building a tuned circuit when you build a pyramid with a slope of 52.66 degrees. This pyramid shape is the three dimensional representation of the carrier wave. The reason it preserves things is the result of the amplified carrier wave created by the shape is amplifying the information that makes up the items you put inside. The first thing we notice about a pyramid is that it puts things back the way they are supposed to exist. It makes things more perfect. The first thing to ask is: How does an object know how to change its condition to a more perfect state than before? Living things use DNA as their template. What about inorganic objects?

The four phenomena observations that tell us the most about what is going on in a pyramid are:

1. The pyramid can be made out of any material and it will still work. Metal pyramids have a greater effect, most likely because they are denser.
2. The pyramid will preserve living objects to live longer.
3. Inorganic objects are affected by the pyramid.
4. The angle and alignment of the pyramid is important to making it work.

When we build a pyramid with the correct angle, we are building a tuned circuit that is a three dimensional representation of the carrier wave. The formula, that I presented earlier and list here, is the three-dimensional representation of the carrier cave. When matter is positioned at this angle it oscillates at the carrier wave frequency. Any object you place inside the pyramid is in essence, having its information amplified. It makes no difference what is put inside the pyramid. The razor blades will stay sharp because the informa¬tion that makes up those microscopic metal crystals will try to remain in their original shapes by trying to move atoms to the areas worn away. The reason any material can be used in building the pyramid is because the one frequency, common to all elements, must be a carrier wave frequency.

The same explanation and conclusion can be applied to the plants and the foods placed under a pyramid. If the plant is growing under a pyramid, it receives amplified information about itself. The stronger the signal, the more energy the plant has. The more the information it receives about itself, the less chance for imperfections in its DNA, which would cause disease, mold, or other organisms to attack it. The increased signals for humans show up as the theta and alpha brain waves, increasing in both frequency and amplification.

This I believe is the formula for the carrier wave and also ties in with e the natural log. The vector angle of energy formed by the carrier wave frequency is 52.66093 degrees. By the way, as a response to those anthropologists who say the Great Pyramid was built by the Egyptians, think about this: Our civilization did not discover e until after Gottfried Liebniz and Sir Isaac Newton invented calculus in the 18th century. There is absolutely no evidence the Egyptians knew about the natural log, not to mention why would they build such a huge building with 2.5 million stones—just to immortalize it?

A final observation: The true builders of the Great Pyramid appear to have had the same philosophy of existence that is expressed in the Theory of Multidimensional Reality.

Chapter Conclusion

The proofs presented demonstrates that there are synchronizing and clocking cycles in the Universe and there is no way to account for them using the matter oriented theory of existence taught today. Only the Theory of Multidimensional Reality can do that. Light is not a particle but information leaving our created reality as a waveform with particles like packets of information. The matter world is the object being created at the speed of light. Time is the rate at which the Diehold transmits our information. The geomagnetic reversal is the result of the main clock cycle reversing every 12068 years. The geomagnetic reversals occur the same time as the ice ages which proves the stars nova also at that time. Dimensions are determined by the amount of energy or information an entity can access and control. Gravity is the result of the "pushing" effect of the information of the planet or star being directed towards the center modulation point. The phantom leaf effect shows that consciousness is transmitted to the same coordinates as the matter body. Finally there appears to be a carrier wave associated with the information that makes up the atoms.

Endnotes

1 The entire proof that our sun Novas is presented in Chapter 3 and 8 in my book *God's Day of Judgment, the Real Causes of Global Warming.*
2 *Ibid.* p.246.
3 Both dates have very high sunspot numbers (over 355 and 342), areas of the sun spots, and other output.
4 God's Day of Judgment, the Real Cuse of Global Warming, 2007 Vector Associates, p.45.
5 Gravitational Mystery Spots of the United States, 1996 Vector Associates,
6 G. Setti, (ed.), *Structure and Evolution of the Galaxies*, (Holland, D. Reidel Publ., 1975).
7 G. O. Abell, *Exploration of the Universe*, 3ed. ed. (Holt, Rinehart & Winston, N.Y., 1975).
8 Thelma Moss and J. Hubacher, Phantom Leaf Effect as revealed through Kirilian Photography (UCLA unpublished, 1975).
9 Bill Kerrell, and Kathy Goggin, The Guide to Pyramid Energy; p.164, (Santa Monica, Pyramid Power V., 1975).
10 *Ibid.*
11 A friend of mine who wrote *The Guide to Pyramid Energy.*

Chapter 3

The Atom and How it is Created

What we know about the Atoms and the Light they Produce

There are 99 different elements, not counting purely laboratory created heavier elements over 99 that do not appear in nature (Appendix B). They all have distinct spectral line frequencies that are used to identify them (Appendix C, an example of the Carbon element, Carbon). Figure 3-1 shows what an atom looks like. The picture on the left shows the original photo, and the right one is the deblurred photo. The deblurred photo shows the electron shells much more clearly than the original. The diameter of the atom shown is 0.5Å^{-1} or 5nm^{-1} and it looks nothing like the Bohr model of the atom or even the theory of neutrons and protons attached.

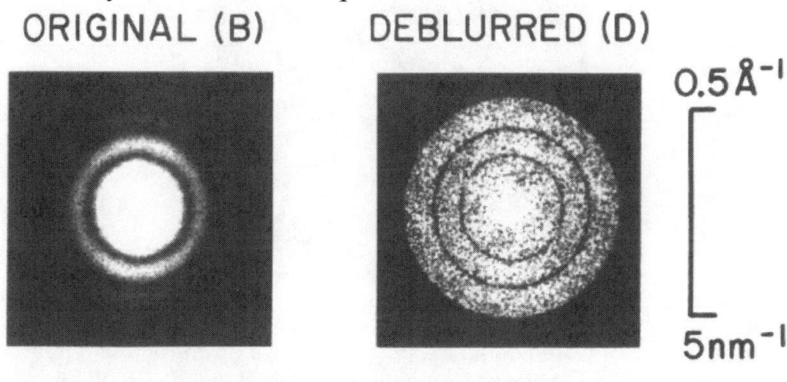

Figure 3-1: Electron Micrograph of heavy atoms on a carbon-foil. Work carried out at the State University of New York, Stony Brook, L.I., N.Y., Electro-Optical Science Center, under the Direction of Prof. G. W. Stroke.

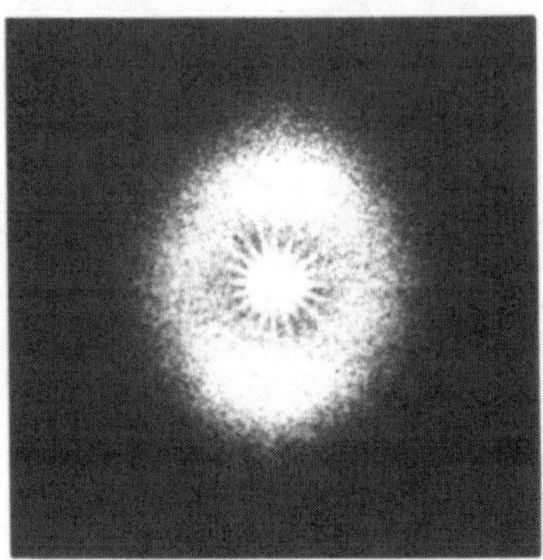

**OPTIMUM
FOCUS
(O)**

Figure 3-2: Electron Micrograph of heavy atoms on a carbon-foil. Work carried out at the State University of New York, Stony Brook, L.I., N.Y., Electro-Optical Science Center, under the Direction of Prof. G. W. Stroke.

When atoms are combined to form molecules they have

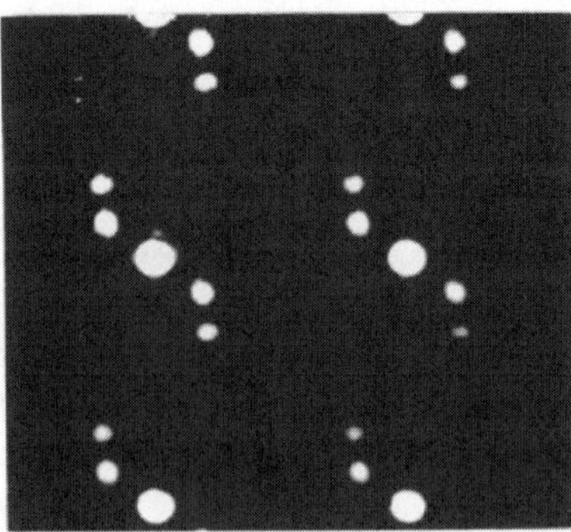

Figure 3-3: Images of atoms in a section of the crystal "magnesium bromide tetrahydrofuran complex," obtained by a scientific team headed by Dr. George W. Stroke and including Dr. M. Halioua, Dr. V. Srinivasan, and Dr. R. Sarma, using the new "X-ray microscopy" opto-digital computing method. The images of the large atoms in the unit shown above are magnesium atoms. The smaller symmetrical pair around each magnesium atom are oxygen atoms, and the still smaller pair farthest away around the magnesium atom, are carbon atoms. The X-dimension of the unit cell shown between magnesium atoms is 9.26Å.

preferred positions to each other, and we do not know why

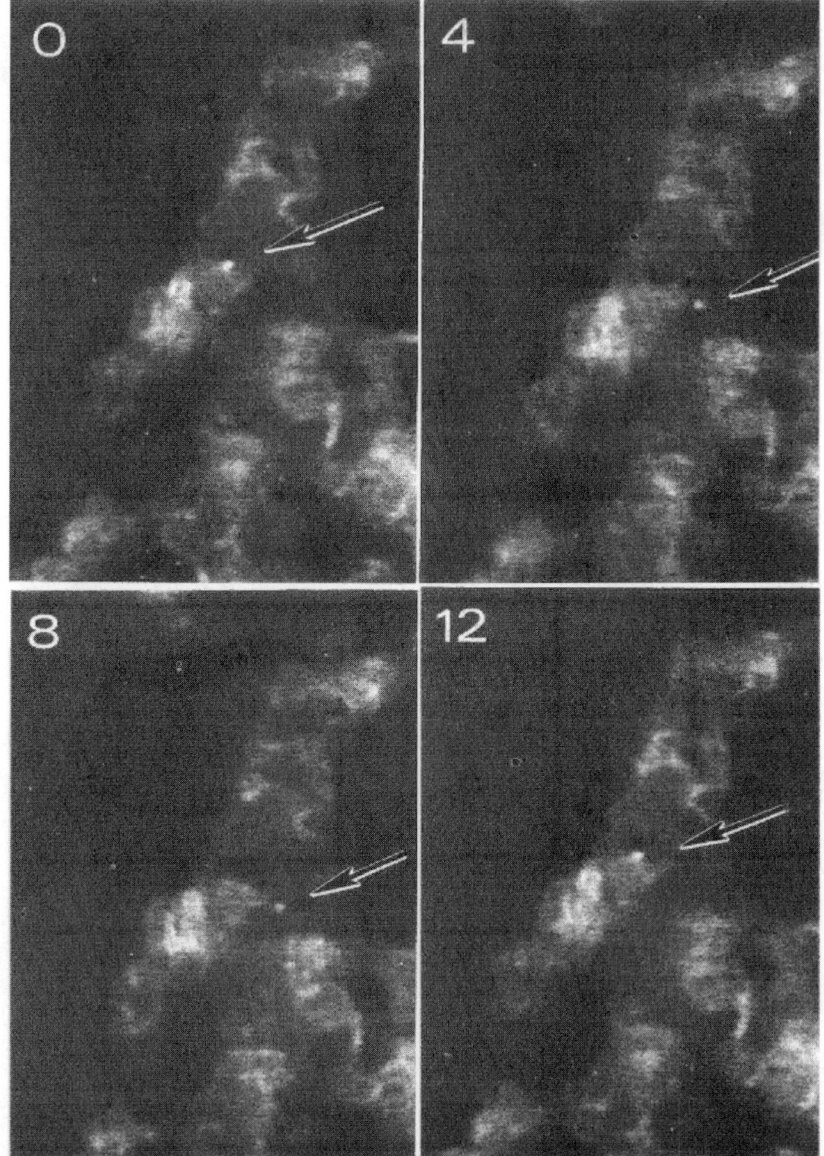

Figure 3-4: The arrow is pointing at a single mercury atom (atomic number 80) at different locations on the carbon film. The pictures were taken every 17 seconds over a 4-minute period, using a 43 keV electron beam. The horizontal field is 125 Å across.

Atomic Jumping Beans

Another phenomenon is that individual atoms are observed moving around in a tight area estimated to be 2-3 Angstrom units and again no one knows why they do it. Figure 3-4 shows two successive pictures of a single mercury atom molecule on a thin film of carbon. The field of view is 125 Å. You will also notice the thin carbon film is also moving.

Subatomic Tracks

When atoms are hit with other atoms or particles, they appear to split apart to form multiple tracks seen in CCD or wire detectors. Mathematicians have made assumptions what some of the myriads of tracks represent and have labeled and grouped them accordingly. Quantum mechanics and the various string theories have attempted to identify, and more importantly, explain this chaos of tracks and what makes them up. As you can see, they are merely trying to explain mathematically the many different tracks by making various assumptions. Figures 3-5 to

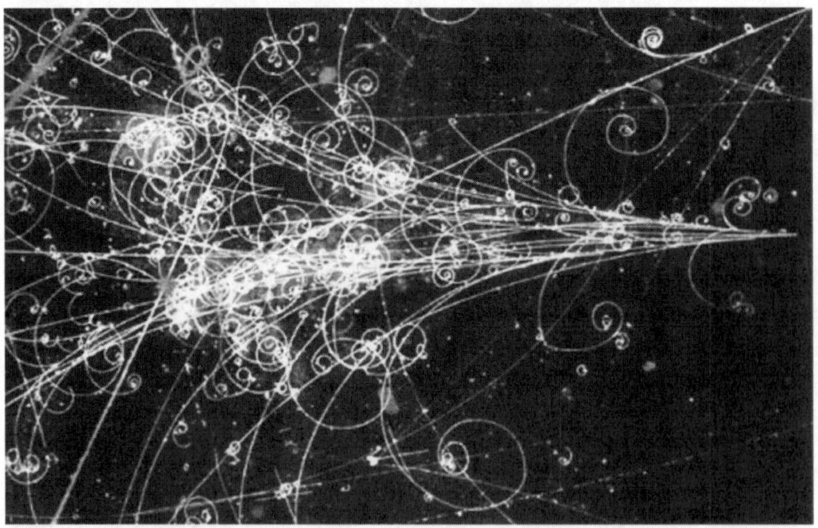

Figure 3-5: Particle tracks from an unknown Bubble Chamber.
3-8 show examples of the subatomic particle tracks.

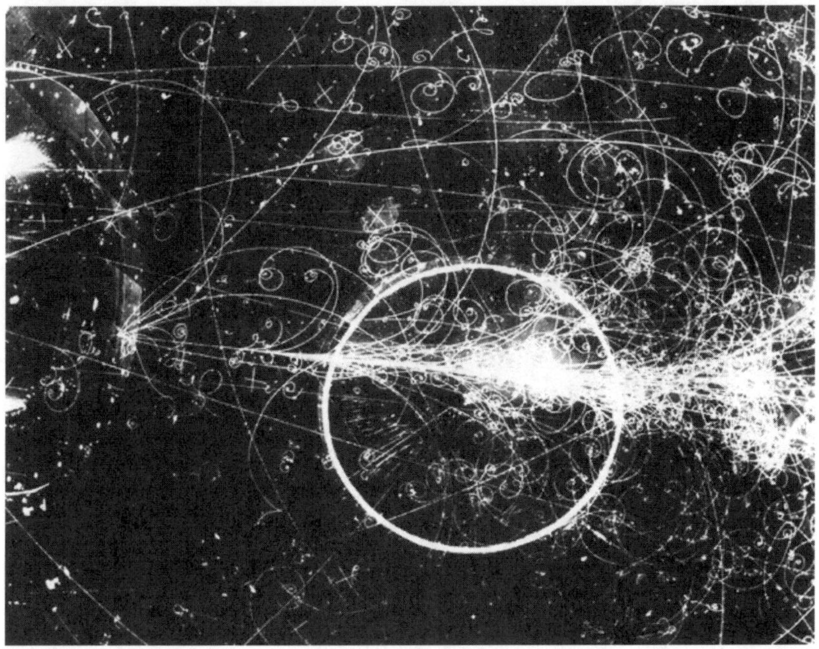

Figure 3-6: From Fermilab Bubble Chamber. The circle and the faint Xs are not the tracks but reference markers.

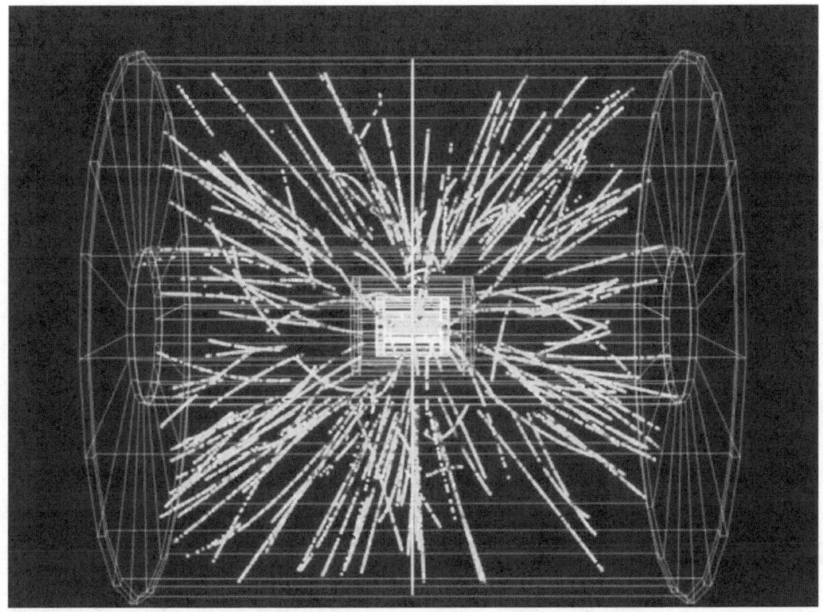

Figure 3-7: The first proton-lead collisions of 2013 send showers of particles through the ALICE detector (Image: CERN)

Figure 3-8: Image of a 7 TeV proton-proton collision in CMS producing more than 100 charged particles (2010). I am told that CERN uses the heavy metal lead as its proton "bullets."

String Theory verses Quantum Mechanics

In standard Quantum Mechanics, particles are considered points moving through space and time, tracing out a line. Particles are said to have position, velocity, mass, electric charge, color (which is the "charge" associated with the strong interaction) or spin. Particles are part of a broader theory called Quantum Field Theory. It provides scientists a theory consistent both with Quantum Mechanics and the Special Theory of Relativity. Quantum Field Theory describes, with great success, three of the four known interactions in nature: electromagnetism, and the strong and weak nuclear forces at the atomic level. Unfortunately the fourth interaction, gravity, does not work into Quantum Field Theory. To apply the rules of Quantum Field Theory to General Relativity you get results, which go to infinities that make no logical sense using the accepted theories of existence. "For instance, the force between two gravitons (the particles that mediate gravitational interactions), becomes infinite, and we do not know how to get rid of infinities to get physically sensible results."[2] However, it does make sense using the Theory of Multidimensional Reality.

There are nine different string theories to date. String Theory replaces the multitude of particles seen in Figures 3-4 and 3-5, with a single fundamental building block called a 'string.' The current science philosophy envisions these vibrating strings to be either closed, in loops, or open, like a violin string. As it moves through time, it traces out a flat sheet if it is an open string (Figure 3-9). You get a closed-ended string (Figure 3-10) when you connect the two opposite end of an open string, it forms a tube.

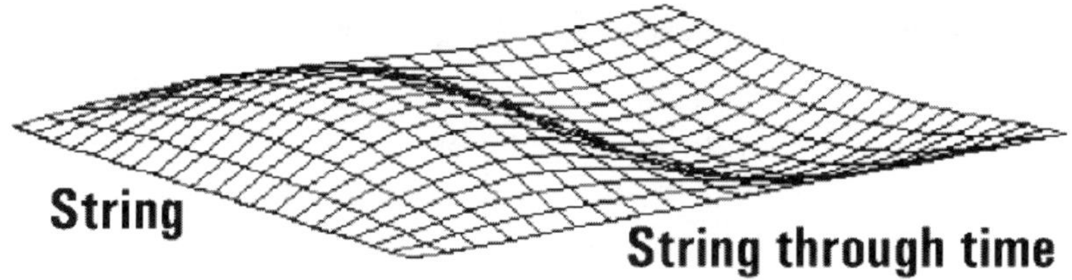

Figure 3-9: Open String as it traces a plane through time.

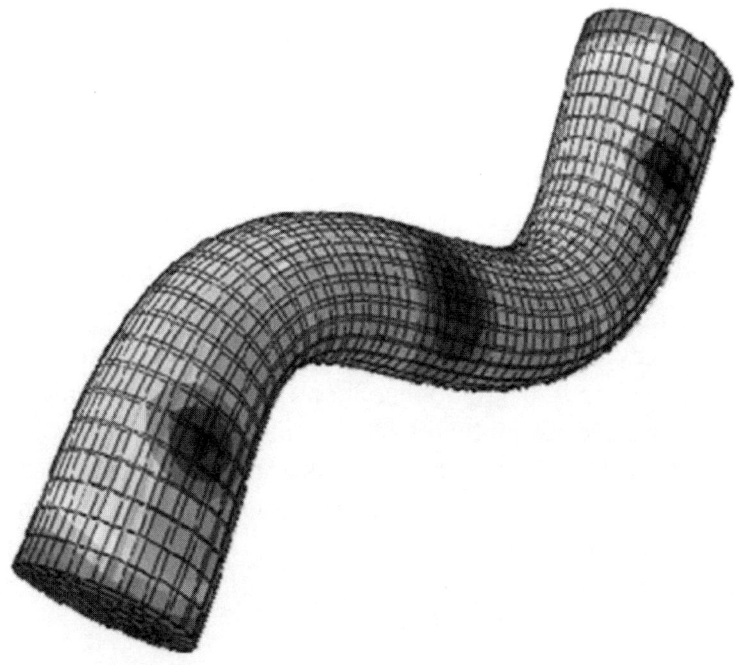

Figure 3-10: Closed end string forming a tube.

The different vibrational modes supposedly create different sub-atomic particles, with different masses or spins. One mode of vibration, or *note*, makes the string appear as an electron, and some as photons. Philosophically their theory does not explain where the frequencies come from that make the strings unique. They theorize that there is a string mode called a graviton, which carries the properties of gravity. String Theory has received so much attention because it allows for gravity in the theory. Unlike Quantum Field Theory, String Theory can account for gravity using the interaction of two gravitons, and there are no infinities to deal with. Gravity in Quantum Field Theory is a factor artificially added to the equation, but in String Theory it has to be there. The great hope of String Theory is that it would unify all known forces and particles in nature into a single 'Theory of Everything.'[3]

The theory of MDR theory helps explain why Quantum Mechanics and both types of string theories are partially correct.

Some Basic Physics Terms

The following four terms are refer to in the next section and important to remember.

Planck's Constant.

Planck's constant is seen as a conversion factor between frequency and energy, especially for photons (traditional explanation for light). It appears in all quantum mechanical equations and is denoted by the letter h and named after the German physicist, Max Planck. Its value is approximately $h = 6.6261 \times 10^{-34}$ Js (Joules). Energy is measured in Joules, often expressed as electron volts (eV) $1\text{eV} = 1.6 \times 10^{-19}$ Js. The energy of a photon (a packet of light) is given by $E = hf$; h is Planck's constant $= 6.6 \times 10^{-34}$ Js and f is frequency.

Planck's Length

Planck's length is considered the basic length resulting from the relationship of three fundamental physical constants: 1. the gravitational constant, 2. the speed of light, and 3.

Planck's reduced constant expressed as ħ. The value is given as 1.616199×10^{-35} meters.

$$\ell_{\mathrm{P}} = \sqrt{\frac{\hbar G}{c^3}} \approx 1.616\ 199(97) \times 10^{-35}\ \mathrm{m}$$

Planck's Time

Planck's time is the time it takes light to travel in a vacuum, the distance of one Planck length. The value is given as 5.39106×10^{-44} seconds (s). It is the relationship between the gravitational constant, the speed of light, and the reduced Planck's constant $(h/2\pi)$.

$$t_{\mathrm{P}} \equiv \sqrt{\frac{\hbar G}{c^5}} \approx 5.39106(32) \times 10^{-44}\ \mathrm{s}$$

All of these values are extremely small units of measure, but we are dealing with the size of atoms.

The Natural Logarithm e^x

The natural logarithm (e^x) shows up in everything. It grows faster than any power of x, as $x \to \infty$. Similarly, e^{-x} decays faster than any power of x^{-1} as $x \to {}^-\infty$. This rapid exponential growth and decay is observed in all physical and chemical reactions; charging and discharging of capacitors, mechanical motion, atomic movement, biological reactions and nuclear reactions. Radioactive elements will decay exponentially over time. Unchecked population growth tends to increase exponentially over time. It is a transcendental number used as the base for natural logarithms. Mathematically, it is defined by the equa-

$$e = \lim_{n \to \infty}\left(1 + \frac{1}{n}\right)^n \cong 2.718281;\ \text{or by}\ e = \lim_{x \to 0}(1 + x)^{1/x} \cong 2.718281$$

tion:

It is represented by the infinite series:

$$e = 1 + \frac{1}{1!} + \frac{1}{2!} + \frac{1}{3!} + \frac{1}{4!} + \ldots + \frac{1}{n!} + \ldots$$

The numerical value for $e = 2.718281$. Graph 3-1 shows a graph plotting the values of the logarithmic function y = *ln* x. This is what a logarithmic function looks like. What is interesting is that there is no philosophical explanation why the natural

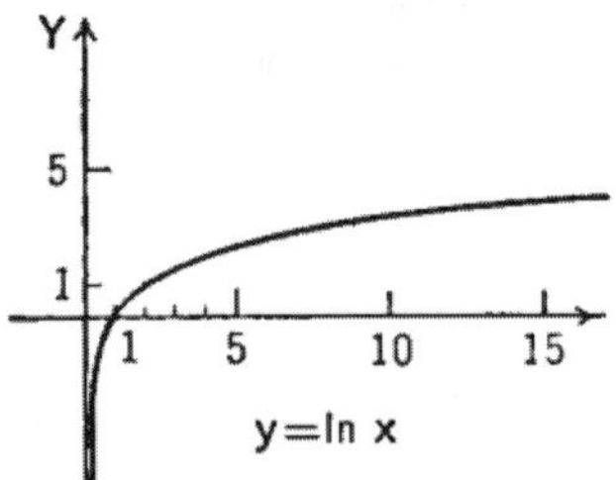

Graph 3-1: Graph plotting the values of the logarithmic function y = *ln* x.

log shows up in our reality. I have not found a philosopher or scientist yet who has explained why *e* shows up in our reality! They have to account for it mathematically as is one of the most important numbers in science, but they cannot explain why.

The Basics of the Model

When writing this Chapter, I was faced with the problem of explaining and applying the Theory of Multidimensional Reality (MDR) to the two competing theories Quantum Mechanics and String Theory. It was important showing the similarities and the differences and how the MDR does a much better job of both modeling the atom as well as explaining how it comes into existence and also leaves this dimension. To accomplish this, I decided to present it first and explain the MDR model of the atom and include how each point of the theory agrees with the other two competing theories.

Graph 3-2 explains five major phenomena within our existence. I will describe each phenomenon below, and what the

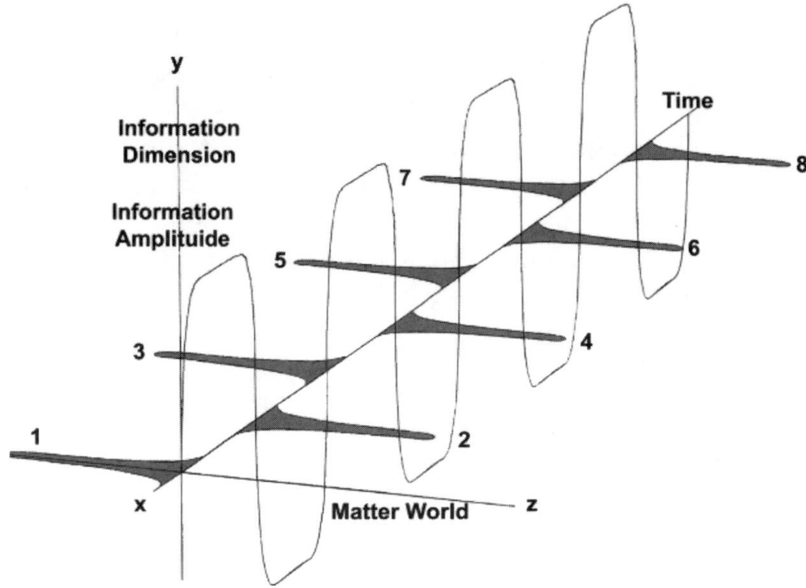

Graph 3-2: The diagram shows the theoretical idea of how a particle travels through time "blinking" in and out of existence in the third dimension. The electromagnetic information of the atom is represented by the Y-axis, the vertical square wave. The horizontal spikes, the Z-axis, represent the electrostatic waveform that is our matter world. They are drawn 90° out-of-phase from each other. Time is shown as the X-axis.

diagram reveals.

First, the Y-axis shows the vertically-oriented modified square wave representing the carrier wave. The Y-axis is the Information Dimension as frequencies directed to a single point in time and space. Its amplitude and frequency vary with each element, and each element consists of many frequencies, but we are only showing one wave in Graph 3-2 namely the carrier wave.

The X-axis represents Time. The speed of light is the rate at which the Diehold processes and transmits all of the information into our Muliverse. In digital terms, we will use Planck's time as the digital rate/speed the Diehold transmits the information of the atoms. Illustrated in Graph 3-2 it would be the time between spikes one to two. In that period of time, it has created what we call a proton and a neutron. Planck's Time would be the time it takes to go from spike one to spike two, measured as 5.39106×10^{-44} seconds.

The Z-axis represents the matter world of the third dimension. It is also Planck's constant of 6.6×10^{-34} Js or expressed as voltage it would be $1eV = 1.6 \times 10^{-19}$ Js. When I first developed this philosophy and diagram, I first assumed it was a spike of energy because it formed 90 degrees out of phase from the information dimension. Then I thought, wait a minute that is also our matter world. This is similar what some scientists in the past thought our atoms were just waveforms. Some of the physicists think now that our reality is a hologram. They just did not know why. Our matter world is created 90 degrees out-of-phase from the information dimension. Planck's length would be between spike two to four or one to three.

Carrier Wave Concept.

The vertical waveform along the Y-axis represents the information dimension (second dimension). The graph shows the waveform as a modified square wave, modified by the natural log, e^x. I will explain why a little later when I explain how and why the natural log appears in our reality. The waveform displayed is like a carrier wave or more accurately described in telecommunication terms as a frequency-division multiplexing wave. This is the technique used by cable and telephone companies to transmit non-overlapping frequency sub-bands. Each band carries different and independent digital information without interfering with other sub-bands. Figure 3-11 graphically illustrates the concept. Also, remember this is just an analogy I use to explain a very complex process that we do not know how it is done to create our reality.

The next question is: What proof do I have of a carrier wave of any type? Near the end of Chapter Two, I covered the known phenomena of a pyramid build to a specific slope angle of 52.66 degrees. You cannot ignore all the phenomena associated with this kind of pyramid just because you cannot explain it using acceptable science theories. The major phenomenon you should notice is that the pyramid can be built out of almost any material, as long as it has mass, it will have an effect on

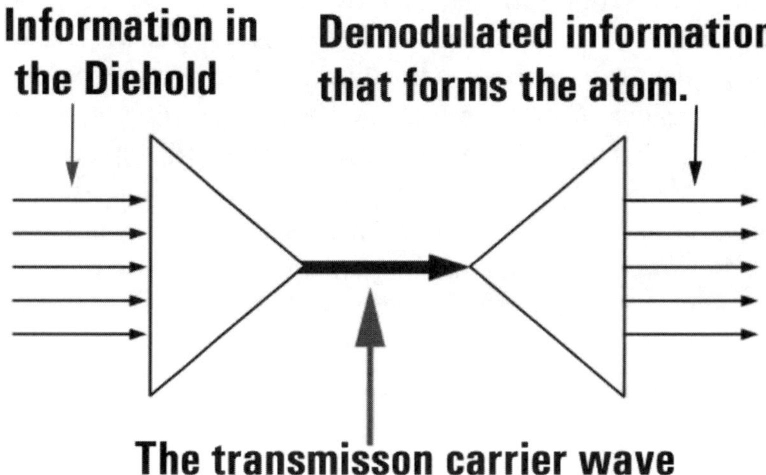

Figure 3-11: A schematic of frequency-division multiplexing applied to the Diehold transmitting information into the third dimension.

objects put inside of it. The fact that any element can be used, and any element will be affected by it, means that there are one or more frequencies that are common to all elements. That is why I came to the conclusion that this shape was a three dimensional representation of a carrier wave common to all elements. I do not see how there could be any other explanation. I believe for life forms there are two carrier waves involved, one for the matter part of the life form (the atoms), and another for the conscious/soul part of the living object. Figures 2-5, 6 and 7 are examples of this effect as seen in the phantom leaf effect.

There is another very important reason I concluded there is a carrier wave involved with the transmission of our information. I covered the reason in Chapter 4 in my earlier book God's Day of Judgment, but An explanation of the conclusion can be found in Appendix D of this book.

Planck's Constant

Graph 3-2 graphically reveals what Planck's constant is, using the Theory of Multidimensional Reality. The energy we call Planck's Constant (6.6×10^{-34} Js) is created when the information frequencies of an atom cross the x-axis. It creates a distinct amount of electrostatic energy, formed 90° out of phase to the collapsing information field. This is our Matter World. What Planck had observed is that our matter world comes into existence in very short pulses of energy. In essence, it blinks in and out of existence constantly, which I depict in Graph 3-2 as spikes one through eight. The odd ones could be looked at as positively charged, and the even ones are analogous to the neutrons, but it is all the same atom, just two different states of its existence. The conclusion we must come to is that our matter world is made up of packets of energy restrained and directed by information from another dimension.

The next challenge: What creates the light from the spectral lines in relation to the model? The light is formed by these spikes of energy, as a result of some of the information leaving this dimension. As previously described, an atom is made up of many frequencies, so we have many of these frequencies going through the same process for the same atom. The resulting light includes the spectral lines of all of those frequencies at much lower frequency than the primary frequency (covered next). The resulting physical atom is the end result of the sums and differences of all its frequencies. The model clearly shows why light exhibits both a wave nature and the particle nature. Light is the result of many packets of information coming into this dimension, creating what we call matter. The matter blinks in and out of our existence. Consequently the resulting light also appears as small packets of energy. Figure 3-12 shows an example of this. Remember the light is the information leaving this time and space. I know this is a difficult concept to grasp but you can look at light as still information receding from us because it is the byproduct of the information that make up the atoms.

The Information for the Elements

We now come to the point of describing the type of digital information that makes up the elements are incorporated into what I believe is the carrier wave. I will be referring to the element table shown in Appendix B. A data sampling of the element carbon is shown in Appendix C.

Figure 3-12: A graphical representation of light taken by scientists at EPFL have succeeded in capturing the first-ever snapshot of this dual behavior. Produced by a research team led by Fabrizio Carbone. Published in Nature Communications March 2, 2015. Simultaneous observation of the quantization and the interference pattern of a plasmonic near-field.

Much can be said about the elements, but I will try to keep it as brief. Of the 99 elements shown in Appendix B, the overwhelming number of spectral lines for all the elements are in the ultraviolet spectrum (UV) with 59.72 percent. Humans can see a range of 350 nanometers (nm) from the low end of 740 nm for IR, up to 390 nm for (UV). Only 32.91% of the elements have spectral lines in our visible light range. The vast majority, 59.72% of the spectral lines are in the UV range above 390 nm. A lot is going on in the high UV range above what we can measure. By my theory each element would be transmitted using a separate carrier wave into the third dimension. That does not mean that each one of those carrier waves have the same frequency. I believe that if Iron has 4,169 spectral lines than all elements above iron should have at least that many, we just do not see them all.

The average number of spectral lines for each element is 439 lines, but the range is enormous ranging from two for Astatine

tine (85) to 4169 lines for Iron (26) and everything in between. There are six elements over 900 spectral lines. The conclusion you have to come to is that we are not looking at all the spectral lines for all the elements. If the information is coming into the third dimension at Planks Time, than we are only detecting much lower harmonics of the primary frequencies. As you look at the table in Appendix B, you will also notice that there is a very wide spread of the spectrum lines for each element. The average spread for all the elements is 16,232 nm, but the range goes from 82 nm for Astatine (85) to Hydrogen (1) with its spectral lines spread over 189,654 nm.

I would conclude that there are element frequencies well above our instrument range for UV. The spectral lines must include the physical structure of the atom, also how it reacts in temperature ranges. How it can combine to other elements and other information we do not even know about. All of this information is somehow strapped onto a carrier wave and transmitted to a point in time and space. Figure 3-16 and 17 shows the toroid shape an atoms path forms.

The Delay in the Propagation of the Information.

The reason Graph 3-2 shows the carrier wave as a modified square wave by slightly curving the corners is in result of the effect of the natural log. It appears in our dimension in various ways including in the circular path the atom takes in the third dimension. Figure 3-16 and 17 shows the toroid shape an atoms path forms.

I briefly mentioned in Chapter Two about the idea of something in the Diehold functioning like our computers' video cards. A video card receives various types of data and combines them into a three-dimensional reality at a slower rate than the CPU of the computer. The natural log ex appears when something is building up or decaying. The MDR explanation is that the natural log appears as a result of information being processed during this "video processing" stage. After that, it enters the second-dimension and finally the created reality of one of the remaining six dimensions. In summary, e^x appears in our dimension because of the delay caused by the processing time of the information from the First Dimension to the Third Dimension. I was asked by readers shortly after my first book Reality Revealed was published: "Where in the sky can I look to see the Diehold?" We do not see the Diehold because it is in another time-space relationship.

Modeling the Atom

The atom is not a bundle of protons and neutrons stuck together as the Bohr model of the atom depicts. That is not what we see in electron micrographs of the atoms. They all look like single points of light. Shown in Figures 3-1 and 2 are micrographs of a single atom. Figures 3-3 show molecules of magnesium bromide. The large atoms are the magnesium atoms (25); the smaller symmetrical pair around each magnesium atom are oxygen atoms (8), and the smaller pair farthest away around the magnesium atom, are carbon atoms (6). Therefore, we can conclude that as the information increases resulting in a heavier element it also become larger in appearance.

We will now take Graph 3-2 and represent the atom as a round three-dimensional shape so you can see what happens as it travels through time. The ball represents a single atom. I

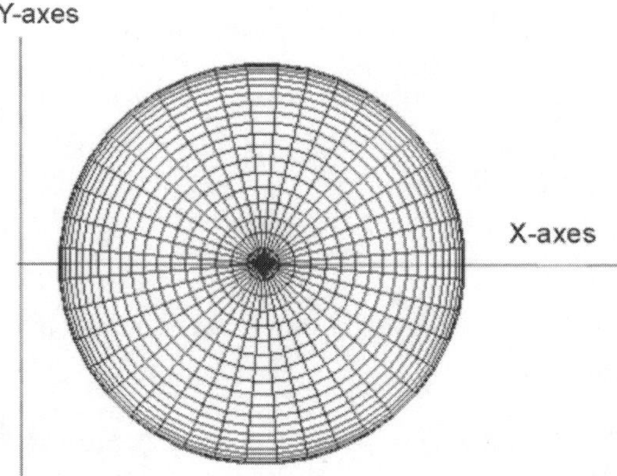

Figure 3-13: Representation of a single atom with its pole pointing along the X-axis.

have drawn it, so the dot is pointing towards the X-axis, which is what happens at the point where its information creates the atom in this dimension.

Now let us follow the pole of the atom as it traces the same waveform along the Y-axis (Figure 4-8) as shown in Graph 3-2. What you see is the pole moving very quickly as it passes through the X-axis. We could rightfully say that it has a polar reversal as the waveform collapses and crosses the X-axis (Figure 3-14). The next point to understand is the atom does not stay in the exact same place, as the Diehold transmits the information to a specific point in time and space. There is a slight difference, as shown in Figures 3-15, with its eight positions—perhaps among thousands of positions as the atom makes one revolution. The shape the circular path forms is a toroid (Figure

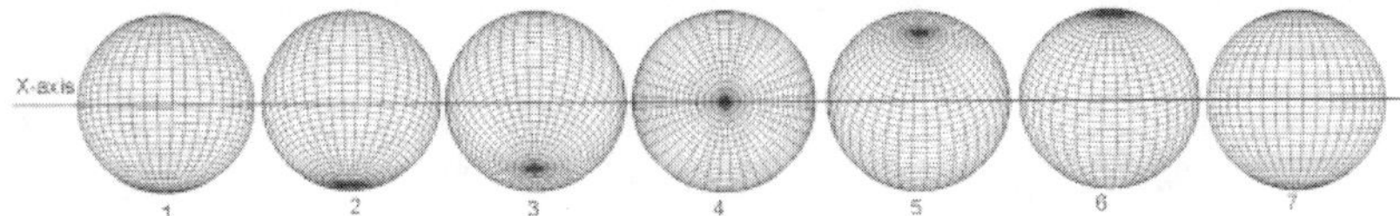

Figure 3-14: The path of the pole as it reverses polarity.

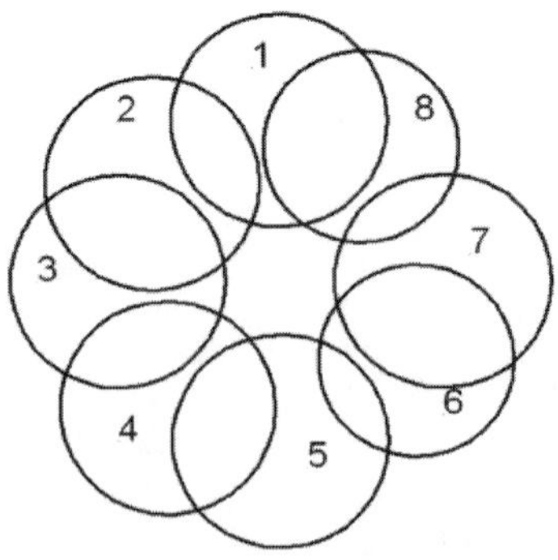

Figure 3-15: Eight positions of an atom as it modulates into the Third Dimension.

3-16 and 17). The atom would have completed two polar reversals by the time it completes one revolution and circles back to position 1 shown in Figure 3-15.

The atom traces out a pattern that is a toroid shape as shown in Figure 3-16. Figure 3-17 shows the path of the polar axis of the atom as it goes around 360 degrees and turns over 360 de-

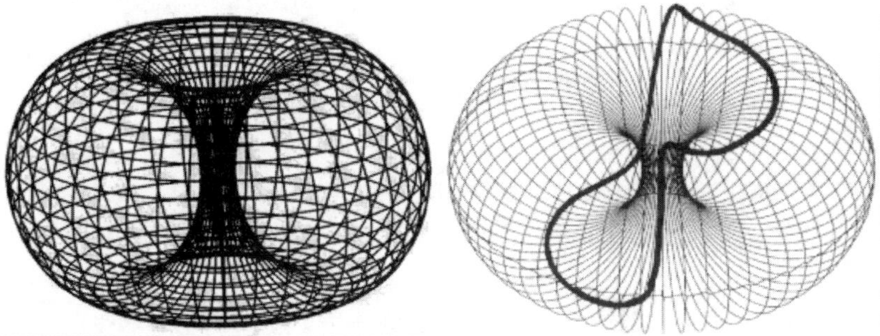

Figure 3-16: A classic toroid shape.
Figure 3-17: The track of the polar axis of the atom.

grees to return to the starting point. A proof of this movement is shown in Figure 3-4 jokingly called atomic jumping beans.

This model is what I used to decipher the 22 letters of the Hebrew alphabet. The Hebrew alphabet is the result of 22 views of this waveform!

The inside diameter of the toroid is determined by how much potential/energy has been imparted to the atom. As more energy is imparted to the atom, the space or hole in the center becomes larger and larger until the information is so unstable it can no longer continue to be modulated into existence to that point in time and space. When that happens, it will quickly demodulate. The different frequencies will fly off in different vectors and intensity depending on the strength of the individual frequency. The theory of MDR concludes that the particle tracks as seen in Figure 3-5 to 8 are created by these different pieces of information hitting other atoms and leaving their paths.

Closed-End String Theory

Closed end strings are merely taking the sheet that is traced out by an open-end vibrating string. Figure 3-10 shows attaching the two opposite sides together forming a tube. When you attached the two ends of the tube together, you get a toroid. That is where string theory is in agreement with MDR. It is also the same regarding the fact that there is a vibrating "string" or frequency involved.

Quantum Mechanics is correct when they state that the atoms emit small packets of energy. Quantum Mechanics and String Theories are wrong in their explanation of light and gravity and subatomic particles.

The Electron

Traditional physics theories envision the electron as a kind of hybrid between a particle and a waveform. In MDR, it is a group wave formed around a center modulation point. Figure 3-1 agrees with that philosophy, so it is a wave in this dimension and electromagnetic or information in the Diehold (the first dimension). Voltage and magnetism in our dimension are both 90 degrees out of phase from what they are in the Diehold.

Black Body Radiation

The term "black body" is a theoretical idea describing an object or area that would absorb all energy and reflect no radiation falling upon it. Its reflectivity would be zero, and its absorbability would be 100 percent. If light were aimed at it, the black body would appear perfectly black and be invisible. The black body could only be detected by what it obscures behind it.

Absolute zero is a theoretical condition when the atom approaches of −273.15°C or −459.67°F on the Fahrenheit scale. At this theoretical point all atomic motion is supposed to stop and no light is given off by the atom. No scientist has ever been able to achieve absolute zero in a lab but they have gotten very close. Also, just because a fine scientist like Robert Boyle came up with the idea of absolute zero does not mean that it is obtainable in our reality and the same goes for black holes, the ultimate in super gravity. It should be considered the ultimate limit of a mathematical theory and not fact until proven by empirical evidence.

The theory of MDR cannot obtain absolute zero because it would mean that time would have to stop and that is impossible in our reality. If no light were given off in any spectrum, it would mean the atom no longer exists in our reality. The diameter of the toroid would get smaller in the center, as the temperature dropped, but would never totally disappear.

Fine Structure

The Fine Structure has been observed using very sensitive spectral analysis equipment, that has shown that many spectral lines are doublets (actually two lines very close together). This phenomenon is believed to be caused by the spin angular momentum of the electron, interacting with the orbital angular momentum of the atom. This explanation is partially correct but can be explained in another way. The Fine Structure is the result of the two phases of the information coming into this dimension, as shown in Graph 3-2. The odd and even spikes would generate a light spectral line slightly different from the other, and therefore would show up as a split spectral line. As stated earlier, it also explains why we think there are separate protons and neutrons. It is just how we see the same waveform but a different phasing of the same information.

Fine Structure Constant

The fine structure constant is a *dimensionless* constant that shows up as the ratio of many different things at the atomic scale. The electromagnetic force that holds atoms together is characterized by the dimensionless Fine Structure Constant = e/hc. The equation incorporates the elementary unit of electric charge e (the electric charge of a proton), the speed of light c, and Planck's constant, h. It incorporates electromagnetism, relativity and quantum mechanics into one equation. The fine structure constant influences all of the sciences. The following quote from the renowned physics professor, the late Robert P. Feynman, sums up the history—and the predicament scientists are faced with—regarding the discovery of the fine structure constant.

There is a most profound and beautiful question associated with the observed coupling constant, e the amplitude for a real electron to emit or absorb a real photon. It is a simple number that has been experimentally determined to be close to -0.08542455. (My physicist friends won't recognize this number, because they like to remember it as the inverse of its square: about 137.03597, with about an uncertainty of about 2 in the last decimal place. It has been a mystery ever since it was discovered, more than fifty years ago, and all good theoretical physicists put this number up on their wall and worry about it.) Immediately you would like to know where this number for a coupling comes from: is it related to pi or perhaps to the base of natural logarithms [2.718281]? Nobody knows. It's one of the greatest damn mysteries of physics: a magic number that comes to us with no understanding by man. You might say the "hand of God" wrote that number, and "we don't know how He pushed his pencil." We know what kind of a dance to do experimentally to measure this number very accurately, but we don't know what kind of dance to do on the computer to make this number come out, without putting it in secretly!

The accepted value is 137 for the Fine Structure Constant. This is supposed to represent the ratio of the strength of the electromagnetic force of the electron, relative to the strong nuclear force that binds neutrons and protons together in the atomic nuclei.

If the science community accepts the Theory of Multidimensional Reality, I would say the physicists have something to worry about regarding this number and the relationships it represents. For one, think the Planck's constant h is the energy a photon (light spectrum) carries with its electromagnetic waveform. The problem is the light represents many frequencies spread over a wide spectrum as has been shown before. Which frequency are you going to choose? The next problem is the value of e which is the charge of a proton. As my model shows the proton is only half the complete waveform and it is also the value of Planck's constant. So what are you really measuring? The speed of light I have no problem with except I theorize it as the rate by which the Diehold transmits the information and that light is information receding from our dimension and we are the objects being created at the speed of light.

Isotopes and Error in the Universe

Scientists have found and identified naturally occurring isotopes for every natural element. They can create isotopes by bombarding atomic nuclei with high-voltage gamma rays, or other particles. Some of these isotopes last only a few milliseconds, and some for many thousands of years. Some of these isotopes release their energy in the form of radiation. Eventually, these isotopes decay down to one of the stable states of the element. The process of decaying isotopes occurs because the Diehold will eventually correct any unstable portions of the Universe.

One might conclude from my theory that the only elements that would appear would be stable, one atomic weight per element. For instance, all of the helium found in our dimension would have an atomic weight of four, or all carbon found would have an atomic weight of twelve. What is known today is that all natural elements have radioactive isotopes.

The general definition of an isotope is: an atom of the same element with the same atomic number, but of a different atomic weight. The isotope is identical in all physical and most chemical properties. The only difference between them is their atomic weight. Some isotopes do behave chemically different. It is almost as if we are dealing with two different elements.

The vast majority of nickel has an atomic weight of 58.7. Why would we find 26.2 percent of nickel with atomic weights of 61 and 3.6 percent at 63, and it will still be nickel? With an atomic weight of 63, it should be copper. Why is it still nickel? The way the MDR theory explains the occurrence of isotopes is by assuming there is error in the Universe. What I did was to create a table with the 99 natural elements. I then listed all of the naturally occurring isotopes for each element. There are also man-made isotopes which last only seconds, and some for many years. I did not take them into account. I only wanted to deal with naturally occurring isotopes that the Diehold normally produces. What I found was that some elements had isotopes that were as much as eight atomic weights greater than the stated average weight, and some were six atomic weights less. Graph 3-3 plots these results.

As you can see from the Graph 3-3, the plot almost fits an

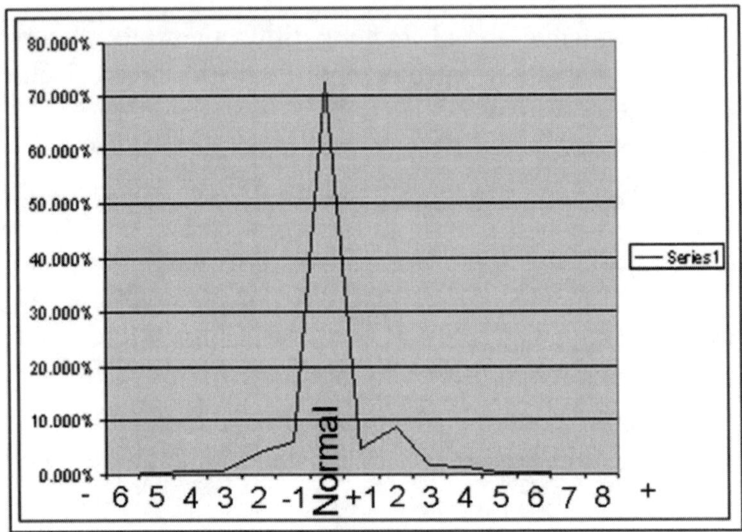

Graph 3-3: Chart is plotting the percentage of isotopes for each natural element. The plot almost follows an exponential curve.

exponential curve, such as the one shown in Graph 3-1. Let us assume that the Diehold is transmitting the information for Carbon. The information consists of many different frequencies. The big question is: How does the Diehold know that it is transmitting the majority of information for Carbon at the correct frequencies? It could be done utilizing the same method used in radio and television transmitters. The de¬vice used is called a *phase detection circuit.* The circuit ensures that the transmitter is producing the correct frequencies. It does this by comparing the correct frequency with any higher or lower frequency produced by the circuit. If the frequencies produced go too high, the phase detector senses it and corrects it by lowering the transmitted frequency. The same thing occurs if the trans¬mitter frequency goes below its acceptable parameters.

The profound point here is this: *Without the error, the transmitter does not know if it is producing the correct frequencies.* I think the same principle holds true for the Diehold. It knows that the majority of signal information is transmitted at the correct frequencies, which in turn modulates into the correct atomic weight for an element. The isotopes are the error factor found in our dimension. If this principle holds true for our Universe, then we should see this error factor show up in everything from childbirths to plant seeds. A certain percentage of life forms will have natural defects, some worse than others, and these defects do in fact show up.

Atomic Fusion and Fission

There is an alternative explanation for atomic fusion or fission events or explosions. Fusion can occur for elements lighter than Iron when relativistic heat is applied to the atoms, so they supposedly combine for heavier elements. Fission is the opposite but for elements only heavier than Iron. When enough kinetic energy is applied to the atoms, they become unstable and supposedly divide and form lighter elements. The slight weight difference in both processes is supposed to create the tremendous heat and explosion.

The MDR explanation is much simpler for both fusion and fission. When you raise the energy level of the atom kinetically in this dimension, you are giving it the equivalent potential in another dimension. When its potential reaches a level too high to be modulated into existence it will demodulate and leave a void or tear in that space and time. The Diehold will not allow that, and it will fill up that space and time with the next most stable element or elements. In the case of fission, it would be two stable elements of average weight. The energy observed is not the difference between the two replacement elements but rather the energy the Diehold can exert to fill a tear in time and space. The same is seen with the energy output of quasars. It can also not be explained from this side of the third dimension.

Déjà Vu

On a personal level, there is a phenomenon that you will all relate to and understand concerning how the natural log reveals itself. All humans have experienced déjà vu multiple times.

Déjà vu happens when you are looking at some event, and all of a sudden you think you dreamt that same, exact event. Usually, the event lasts for less than a second. No action on your part can change the outcome of the event—it just happens. Our usual reaction is we think we must have dreamt the event and did not recall the dream until now.

If our existence emanates from the Diehold, in another dimension, then our thinking processes are also occurring there. The matter part of our brain is like a receiver for this information. What is happening with déjà vu is that the information that makes up what we are doing and seeing is perceived twice. The first signal is perceived directly from the Diehold, the second signal through our physical senses. It is processed by our brain, like audio feedback. The signal is sent through the circuit twice. I do not know why this phenomenon happens in the Diehold, or why some people have it more often than others. The delay in the signal is an example of e^x, the natural log and what is going on in the video processing section of the Diehold.

Endnotes

1 Stroke, George W. Et Al. IEEE Spectrum, 9(12): 24-41 December 1972.
2 Cambridge University Quantum Theory web site: http://www.damtp. cam.ac.uk/user/gr/public/qg_ss.html.
3 The latest evolutionary version of String Theory is called M-Theory that calls for 11 dimensions and combines a little from each of the five other String Theories. M-Theory is more convoluted than the others and I will not cover it here.

Chapter **4**

Why was the Universe Created?

In the Beginning . . .

In the Preface I listed the three most important questions in the Universe. They are: "Why was the Universe created?" "If God created the Universe, then who created God?" " Why was man created and his place in the Universe?" These have to be the most difficult questions in the entire Universe to answer—difficult, but not impossible. If a theory of existence can answer these three questions, it must be the secret of the Universe and how it really works. This chapter will answer these three questions.

Summary of what we think we know:

The previous chapters should have proven to you that the Universe is the product of information. We are in a computer generated reality transmitted from another time and space and have an operating system which we call God. This "computer" which I have named "The Diehold," holds all the information of our Universe including the Operating System. Our Diehold has to be constructed out of matter, otherwise information could not be contained within it, to be processed, and transmitted to create our existence. If it is constructed out of matter then the Theory of Multidimensional Reality would conclude it is transmitted into existence from another Diehold—not ours but one above ours. I think you can see where I am going with this. If I have presented enough proof that the Universe is the product of information, then it logically leads you to the conclusion that a very highly advanced civilization must have built our Diehold and turned it on. I will refer to them as the "Builders." The action of turning on our Diehold created the Big Bang. Something cannot come from nothing but it can be created from information from outside the created reality. However, that still does not tell us why they built our Diehold and man's purpose, in the scheme of things. To answer that, we must delve deeper.

The Answer is in Genesis

In order to unravel this great mystery of what Man is in relation to the Universe we must start at the beginning, Genesis Chapter 2:8: "And the Lord God planted a garden eastward, in Eden; and there He put the man whom He had formed." The Hebrew word for formed can also mean pattern. So what this verse reveals is that God created a pattern that represented man. In other words, He created a program that represented man, which the Diehold creates into existence. A program cannot "die" in a computer unless it is destroyed or turned off by the Operating System. Your soul is the program and it does not "die" after your physical body (receptacle) stops functioning.

Our next clue is found in Geneses 2:17: "But of the tree of the knowledge of good and evil, thou shalt not eat of it; for in the day that thou eateth thereof thou shalt surely die." The verse implies that Adam's (Man's) previous state of existence was to live forever, but if he entered some device, allegorically represented as the "tree of knowledge of good and evil," Man would then, be mortal. The same theme is again repeated in Chapter 3:3 "But of the fruit of the tree which is in the midst of the garden, God hath said: 'Ye shall not eat of it, neither shall ye touch it, lest ye die'." The message is loud and clear, these people were in a state of existence that they would live forever. In *God's Day of Judgment*, Chapter 4 goes into great detail the Garden of Eden story and what Adam and Eve did. For now I will say that they were sixth dimensional beings that made a conscious decision to de-evolve into fifth dimensional beings, but the Biblical story does not tell us why they did it, and that is the secret you are about to discover.

An Advanced Civilization—The Builders

What kind of civilization could build a computer to replicate their universe? It would have had to be either a highly advanced fifth dimensional beings or six dimensional beings who have been around for more time than I am willing to guess. Why would an advanced civilization want to build a replica of their own Diehold that they know they are in? There are two reasons.

The Simple Reason

The first reason I thought of was the obvious one. The principle of the experiment is: Do you get out the same program that you programmed into it? There must be a mechanism within the Diehold, which enables the Operating System to report to the "Builders" of the Diehold. I thought of a simple experiment that would prove whether the Operating System permeated all levels of existence. Allow me to explain what I mean. Imagine you are standing between two parallel mirrors. When you look at one, you see yourself reflected an infinite number of times in the two mirrors. Another analogy is taking a computer program and copying it onto a compact disk (CD), loading it into another computer, and than copying the same program onto another CD and loading it into another computer. Let us say that you copy it over a million times. The millionth copy of the program will be the same as the first one. Do we care what "copy" we have? No. Do we care if we have the first copy or the 300th or the 300,000th copy? No. The experiment would be to see how long it would take for intelligent life, in your Diehold, to arrive at the correct programming for the Operating System of the Universe and to feed the result back to you in its entirety. If the Builders received back the same Operating System program they put in, that would prove the Operating System permeates all levels of existence. Most likely the spaces (distances) between the atoms, form the patterns for the program, and it permeates through all levels of existence. This could be related to the carrier wave, which could be carrying the Operating System.

What is profound about this experiment is that it proves the Universe started with a thought form and that started it all. In other words, the Ultimate God started it all with a thought form with the correct pattern.

Our Universe is thought to be about 18 billion years old, dating from the Big Bang, when it is supposed to have started. When I came up with the first reason for creating the Universe, I wondered: "How much time would it take for the Builders of our Diehold to get the answer from their Diehold? I figured it would take probably no more than 4.5 billion years of our time. I came up with this number by assuming that planets would form within half a billion years of the Big Bang, and it would take the rest of the time for a planet to create the right conditions to produce intelligent life. Than it was just a matter of the odds of a planet developing an advanced intelligent civilization. Of the many trillions of solar systems in the Universe, what percentage might produce life which would evolve to the fifth or sixth-dimensional level? I figured there would be quite a few. So, for this discussion, let us say the Builders of our Diehold would have gotten their first answer after 4.5 billion years of our time.

Now let us say their Diehold has a clock cycle of 100 megahertz, which is rather slow by today's standards. Remember that one clock cycle in our Diehold equals 12,068 years of our time. Let us now do a little calculating to see how long it would take the Builders to receive their answer, in their time. Let us first convert our Earth time into clock cycles in the Diehold. We divide 4.5 billion years by 12,068 to get 372,887 clock cycles. Next, we convert to the Builders time standard by dividing 372,887 by 100,000,000, and we get .00373 seconds. Even if we assume it would take 18 billion years for a universe to produce its first sixth-dimensional beings, it still converts down to only .0149 seconds. If the speed of the computer goes up, it will take less time.

I was rather taken aback after realizing that it would take so little time to get the answer. It would be like turning on your computer and receiving the answer before you could take your finger away from the ON button. It did not make any sense. Would any civilization go to the expense of building a Diehold for an answer they would get back almost immediately? I don't think so.

In summary, the first test or experiment is an important one but it is not the primary reason the "Builders" built our Diehold.

Man

Man is very special to the Diehold. In fact, I have to conclude it was designed for man to evolve in. When I say man, I mean fourth, fifth and six dimensional intelligent beings. Nevertheless, man is a special program within the Diehold, with a special function, which I will explain later.

The Ultimate Question: Why was the Universe Created?

I came up with the answer in the year 2000 and it left me emotionally flat. So I warn you, you may not like the answer but after you think about it, you will agree it is the only answer.

I have found that people and governments act on things for their own selfish reasons. There is nothing wrong with self-interest. After all, that is what competition is all about, which creates progress, not stagnation. Without progress and inventions, you wind up with a bronze age civilization with only thirty years left before the polar reversal, and they are ignorant of the events which are going to happen, and do not have the technology to save themselves. Our civilization fits into that category also.

Therefore, there has to be another reason why our Diehold/ Universe was built. The answer was written in the Genesis story of Adam and Eve. I previously showed that the civilization Adam and Eve belonged to was a sixth-dimensional society where people lived forever because they perceived all of their reality from within the Diehold without a physical body. The Torah and I mean that literally—these people exist forever. After all, if you are a computer program in the Diehold, there is no reason why you would die, unless the Operating System decided to destroy your information, for whatever reason.

Now, since I have established that, I want you to think about living forever and remembering everything you have ever done, seen, lived, read, loved, hated, etc. Now imagine that you have lived and reincarnated as different people, not hundreds or thousands of times, but millions-upon-millions of times. At some point you would have seen every imaginable life form. You would have seen every form of government, religion, science, technology, life style, good and bad. I believe you can see where I am heading with this. At some point in your existence the repetition and resulting boredom would become so unbearable that you would do anything to forget. To forget everything you did in previous lives, who you knew, what you did, even the enjoyable events. It would be so painful to remember all of it, that you could be driven to madness. Many people say they wish they could remember what they were and did in a previous lifetime, but I believe that may be a mistake. What I believe now is that it is a blessing to forget. The reason why Adam and Eve decided to devolve to fifth-dimensional beings was so they could forget and live a life free of their past memories. They wanted to be human again. God was not tremendously upset with them. He just honestly informed them what existence would be like after they took this lower form of life. I am beginning to believe that most sixth-dimensional beings devolve for the same reason. I would not term it entertainment for them, but more like "what you have to do to keep your sanity."

The Answer

To return to the Builders of our Diehold, I believe their civilization came to a critical point in their existence. They knew that they are also in a Diehold which creates their existence. They also knew that the Diehold they built (our Universe) is teeming with life and advanced civilizations within it, who are also building Diehold's to pose the same eternal question. That question is: Will some sixth dimensional being inside their Diehold turn it off, and how long would it take?

Yes, I believe that when these computers are built there is a mechanism for a sixth dimensional being to turn it off. It might take the form of an octahedron-shaped structure, which mirrors the actual Diehold computer, but is a created image with an equivalent software off switch.

The moral question for any advanced civilization is: Do they have the moral right to shut it down? I take the moral position that we do not have that right. The Universe is teeming with life. That is why the advanced civilization that created the letters on the two tablets Moses retrieved from the family cave is derived from a model of the carbon atom because all life forms are carbon based. That is also why they only had 22 letters. Once you realize that one Diehold eventually spawns untold billions of additional Dieholds within it, you realize that when a Diehold is turned on it is literally an explosion of life going on inside.

I am sure many of you will be shocked at my conclusion why our Universe/Diehold was created. Once you accept that we are in a created reality, we are created from information and the Torah clues says some form of Man can live forever, there is no other logical answer to the question.

Why did God become irritated with Adam and Eve? I believe that the Operating System/God is pro¬grammed to encourage upward evolution, so as to test the question will someone inside the Diehold, turn it off? That is why He gave us all the Laws of conduct in the Torah. He wants Man to evolve to an eventual understanding of how the Universe really works and what His relationship is to it.

How about Us?

We are only fourth-dimensional beings so we have a long way to go before we need to worry about this type of decision. I now believe that many of us have devolved many times, and have chosen to become fourth-dimensional beings so we could spend a much longer time evolving upwards, thus forgetting longer. I believe God gave us the Ten Commandments, and many other rules of conduct, because He knows the best way to live and not make life a living hell. You cannot argue with an Entity that has seen and controls it all.

Conclusion

I have covered many subjects and philosophical issues in this book. I have delivered what I promised you in the Preface. You now all know there is a God, and what his relationship is to the Universe He creates. He may not be what you envisioned, or what you were taught, but this is how the Universe really works.

Endnotes

1   I created the name Diehold in 1975 because there was no other name in the English language, or others that described a computer which holds all of existence within it.
2   I have capitalized "Operating System" because the term represents God, and traditionally, all references to God are capitalized.

Appendix A

Table of Stars, Open Clusters and Globular Clusters

Table of Stars in the Milky Way Galaxy

I include the following table because of its importance and so other researchers could follow in my path.

The following table lists the Open Clusters (O) and Globular Clusters (G) found in our Milky Way Galaxy. I did not list the stars closer than 9,400 light years even though they are part of my complete database. An Open Cluster is a loose collection of stars in a small area of space. They usually number from a few hundred to a few thousand stars. A Globular Cluster is a tightly packed collection of stars of all sizes. A globular cluster contains hundreds of thousands or millions of stars. There are over 100 globular clusters in our galaxy.

The following table is dominated by Open Clusters up to a distance of 13,400 light years (LY). After that there is a mixture of Open Clusters and Globular Clusters. After 17,000 LY the database is dominated by Globular Clusters. This is probably because it was difficult for astronomers to reconcile single-light (*i.e.,* of the smaller open clusters) sources so far away.

Astronomers measure stellar distances in units of measure called parsecs. Each parsec is equal to the distance light travels in 3.262 years. You can see we are dealing with some very great distances and of course some subjective decisions astronomers have to make to determine these distances. What is interesting is that the six different astronomers who determined the distances for the stars at the beginning of the six blank periods had their calculations come out to reveal the 12,068 number. They of course, had no idea of the importance of that number. There for the subjective part of their calculations were not influenced.

To make the stellar graph (Graph 2-1) display the blank

periods correctly, I added two records with a value of zero just after the star cluster and just before the next star cluster. The data comes from several sources but predominantly from *Sky Publishing's* 1981 directory of stars call Sky Atlas 2000, of Open Clusters and Globular Clusters. The star systems that were close to the time/distance periods I checked with other astronomical sources and picked the one that best fit my model. I know this is not what you would normally do in research, even though it is done all the time. It is like cherry-picking your results but remember I already knew there was a know error factor of 2.5% to 10%. The important thing to realize is that there is a degree of subjective interpretation of the raw results. Moreover, as I have said before, I think God influences man by putting ideas in our minds to lead us to His desired results. I only started looking for other distance results after I started seeing a pattern. The important constellations are footnoted as to their sources. At no time did I alter anyone's distance figures. All I did was convert their findings from parsecs into light years and graph them.

Constellation	Name	Type of Stars	Quant.	Dist./PC	Dist LY
HOR	AM-1		1		0
MON	NGC 2324	O	70	2900	9458
OPH	PAL 6	G	50000	2900	9458
CEN	CR 272	O	40	2900	9458
PUP	NGC 2483	O	30	2900	9458
CAS	NGC 103	O	30	3000	9785
CAS	NGC 609	O	25	3100	10111
SGR	NGC 6656	G	100000	3100	10111
CEP	NGC 7510	O	60	3160	10306
NOR	NGC 6031	O	20	3200	10437
PER	BERK 68	O	60	3200	10437
PYX	NGC 2818	O	40	3200	10437
CAS	NGC 7790	O	40	3200	10437
CAS	BERK 65	O	20	3300	10763
CAR	NGC 3324	O	25	3300	10763
CAS	IC 166	O	120	3300	10763
PUP	NGC 2467	O	50	3400	11089
GEM	NGC 2266	O	50	3400	11089
MON	NGC 2236	O	50	3400	11089
CAR	NGC 3603	O	30	3500	11415

Constellation	Name	Type of Stars	Quant.	Dist./PC	Dist LY
CYG	NGC 7067	O	20	3500	11415
CEP	NGC 7380 [1]	O	40	3600	11741
1 ST	**DARK AREA**		0		11742
1 ST	**END DARK AREA**		0		13045
CRU	RU 97 [2]	O	20	4000	13046
AUR	NGC 1893 [3]	O	60	4000	13046
AUR	BERK 19	O	40	4000	13046
OPH	NGC 6366 [4]	G	100000	4000	13046
SGR	NGC 6440	G	50000	4100	13372
PUP	RU 32	O	30	4100	13372
SGE	NGC 6838	G	50000	4100	13372
AUR	KING 8	O	30	4150	13535
CAR	PISMIS 17	O	25	4200	13698
CRU	HOGG 15	O	15	4200	13698
PAV	NGC 6752	G	100000	4300	14024
AQL	NGC 6760	G	40000	4300	14024
CIR	PISMIS 20	O	25	4400	14351
OPH	NGC 6254	G	100000	4400	14351
PUP	RU 55	O	12	4400	14351
ORI	NGC 2141	O	100	4400	14351
CMA	NGC 2204	O	80	4450	14514
TUC	NGC 104	G	50000	4600	15003
SGR	NGC 6544	G	100000	4600	15003
CMA	NGC 2243	O	100	4600	15003
MUS	NGC 4372	G	100000	4900	15981
GEM	NGC 2158	O	25	4900	15981
SGR	NGC 6603	O	50	4906	16001
MON	BIUR 10	O	20	5000	16308
VEL	NGC 3201	G	100000	5000	16308
CAR	WESTR 2	O	12	5000	16308
LYR	NGC 6791	O	300	5100	16634
CEN	NGC 5139	G	1000000	5200	16960
MON	DO 25	O	50	5200	16960
SGR	NGC 6809	G	100000	5200	16960
OPH	NGC 63047	G	40000	5400	17612
ARA	NGC 6352	G	50000	5400	17612
MUS	NGC 4833	G	75000	5500	17938
OPH	NGC 6218	G	100000	5500	17938
OPH	NGC 6171	G	100000	5900	19243
AUR	BASEL 4	O	15	5900	19243
SGR	NGC 6553	G	100000	5900	19243
OPH	NGC 6266	G	50000	6000	19569
SGR	NGC 6626	G	100000	6100	19895
SGR	NGC 6642	G	40000	6200	20221
SGR	NGC 6522	G	40000	6400	20874
PUP	RU 44	O	40	6600	21526
OPH	NGC 6401	G	40000	6800	22178
OPH	NGC 6355	G	40000	6800	22178
CRA	NGC 6541	G	100000	6900	22504
OPH	NGC 6333	G	100000	6900	22504
ARA	NGC 6362	G	100000	7100	23157

Constellation	Name	Type of Stars	Quant.	Dist./PC	Dist LY
SCO	NGC 6453	G	20000	7100	23157
HER	NGC 6205	G	100000	7200	23483
OPH	NGC 6293 [5]	G	50000	7300	23809
2ND	**START DARK AREA**		0		23810
2ND	**END DARK AREA**		0		25439
SGR	NGC 6528 [6]	G	20000	7800	25440
SCT	NGC 6712	G	50000	7800	25440
HER	NGC 6341	G	100000	7800	25440
LUP	NGC 5927	G	100000	7800	25440
SGR	NGC 6712	G	50000	7800	25440
SGR	NGC 6638	G	40000	8000	26092
SCO	NGC 6144	G	70000	8100	26418
SGR	NGC 6569	G	40000	8200	26744
CAP	NGC 7099	G	100000	8200	26744
SCL	NGC 288	G	100000	8300	27070
SCO	NGC 6093	G	100000	8300	27070
SGR	NGC 6624	G	40000	8500	27723
SCO	NGC 6380	G	20000	8700	28375
OPH	NGC 6517	G	40000	8700	28375
SGR	NGC 6723	G	100000	8700	28375
SCO	NGC 6139	G	50000	8900	29027
CEN	NGC 5286	G	100000	8900	29027
OPH	NGC 6287	G	30000	9000	29354
MON	BIUR 8	O	70	9000	29354
SGE	PAL 10	G	30000	9000	29354
SCO	NGC 6496	G	50000	9000	29354
TUC	NEC 362	G	100000	9000	29354
SCO	LILLER 1	G	20000	9000	29354
CAR	NGC 2808	G	100000	9200	30006
SGR	NGC 6558	G	20000	9200	30006
PEG	NGC 7078	G	100000	9400	30658
LYR	NGC 6779	G	40000	9500	30984
SER	NGC 5904	G	10000	9600	31310
HYA	NGC 4590	G	100000	9600	31310
NOR	NGC 5946	G	50000	9600	31310
CVN	NGC 5272	G	100000	9900	32289
SGR	UKS1751-241	G	30000	10000	32615
OPH	NGC 6235	G	50000	10000	32615
OPH	NGC 6284	G	50000	10000	32615
SCO	GRINDLAY 1	G	30000	10000	32615
LUP	NGC 5986	G	100000	10200	33267
OPH	NGC 6402	G	100000	10200	33267
SCO	NGC 6441	G	50000	10300	33593
SGR	NGC 6637	G	50000	10300	33593
OPH	NGC 6273	G	100000	10600	34572
SGR	NGC 6681	G	50000	10800	35224
COL	NGC 1851	G	100000	10800	35224
AQL	PAL 11	G	20000	11000	35877
SER	NGC 6535 [7]	G	20000	11000	35877
3RD	**START DARK AREA**		0		35878
3RD	**END DARK AREA**		0		36854

Constellation	Name	Type of Stars	Quant.	Dist./PC	Dist LY
AQR	NGC 7089 [8]	G	100000	11300	36855
OPH	NGC 6316	G	30000	12000	39138
ORI	BERK 21	O	40	12000	39138
LIB	NGC 5897	G	100000	12100	39464
PUP	NGC 2298	G	100000	12300	40116
APS	NGC 6101	G	100000	12300	40116
SER	IC 1276	G	50000	12900	42073
LEP	NGC 1904	G	100000	13300	43378
HOR	NGC 1261	G	75000	13400	43704
SCO	NGC 6388	G	50000	14500	47292
BOO	NGC 5466	G	100000	14500	47292
DEL	NGC 6934 [9]	G	40000	14700	47944
4TH	**START DARK AREA**		0		47945
4TH	**END DARK AREA**		0		48922
SGR	NGC 6652 [10]	G	20000	15000	48923
TEL	NGC 6584	G	50000	15000	48923
OPH	NGC 6342	G	20000	15000	48923
COM	NGC 5053	G	100000	15200	49575
SGR	NGC 6717	G	20000	16000	52184
OPH	NGC 6426	G	30000	16000	52184
OPH	NGC 6356	G	60000	17200	56098
COM	NGC 5024	G	50000	17200	56098
AQR	NGC 6981	G	40000	17300	56424
COM	NGC 4147	G	50000	17500	57076
SGR	NGC 6864 [11]	G	50000	18200	59359
5TH	**START DARK AREA**		0		59360
5TH	**END DARK AREA**		0		60663
APS	IC 4499 [12]	G	75000	18600	60664
CAP	PAL 12	G	20000	19000	61969
OPH	NGC 6325	G	30000	19400	63273
SER	PAL 5	G	70000	21400	69796
SGR	NGC 6715	G	100000	21500	70122
VIR	NGC 5634	G	150000	21600	70448
AQR	NGC 7492 [13]	G	50000	21900	71427
6TH	**START DARK AREA**		0		71428
6TH	**END DARK AREA**		0		77297
LUP	NGC 5824 [14]	G	50000	23700	77298
PEG	PAL 13	G	20000	24400	79581
SGR	PAL 8	G	40000	30800	100454
HER	NGC 6229	G	30000	31200	101759
HYA	NGC 5694	G	30000	32300	105346
DEL	NGC 7006	G	10000	34700	113174
AUR	PAL 2	G	25000	35000	114153
CEP	PAL 1	G	20000	46000	150029
OPH	NGC PAL 15	G	30000	70000	228305
HER	PAL 14	G	20000	75000	244613
LYN	NGC 2419	G	35000	93100	303646
UMA	PAL 4	G	25000	93300	304298
SEX	PAL 3	G	30000	96000	313104

Endnotes

1 Wil Tirion, Sky Atlas 2000, (Cambridge, MA, Sky Publishing Corporation, 1982), p. 284.
2 Ibid. p. 280.
3 Ibid. p. 274.
4 Ibid. p. 294.
5 Ibid. p. 294.
6 Michael Rowan-Robinson, The Cosmological Distance Ladder: Distance and Time in the Universe, (New York: W. H. Freeman and Co, 1985), p. 321.
7 Wil Tirion, Sky Atlas 2000, (Cambridge, MA, Sky Publishing Corporation, 1982), p. 294.
8 Ibid. p. 294.
9 Ibid. p. 294.
10 Ibid. p. 294.
11 Ibid. p. 294.
12 Ibid. p. 293.
13 Ibid. p. 294.
14 Ibid. p. 293.